Publishing Information

ISBN: 978-1-60888-314-1

Bibliographic Key Phrases

Codes; Code Language; Lisbon Telegraph Conference 1908; John C. Hartfield; Artificial Words; Roots and Terminals; Official Vocabulary; Morse Code; Dalgety & Company, Ltd; Signal Intelligence

Publisher's Notes

In a world grappling with cybersecurity and data privacy, understanding the history of secure communications is more crucial than ever. *John C. Hartfield's 'Lisbonian' Selection of Roots and Terminals offers* a fascinating glimpse into the intricate world of telegraph codes, a vital tool for confidential communication in the early 20th century. This document delves into the meticulous construction of a code language, featuring 21,323 five-letter words, which translates to an astounding 454,670,329 ten-letter words designed for secure and efficient transmission via telegraph. Hartfield's work directly addresses the challenges and debates surrounding the Lisbon Telegraph Conference of 1908, which sought to regulate the use of artificial words in telegraphic communication. This document provides unique insights into the methodologies employed by Hartfield, including his ingenious "Detectors," which allowed users to decipher and correct errors in transmission. It also features fascinating testimonials from prominent companies like Dalgety & Company, Ltd., highlighting the effectiveness and reliability of Hartfield's code system in real-world scenarios. "John C. Hartfield's 'Lisbonian' Selection of Roots and Terminals" is a must-read for researchers of communications history, cryptography enthusiasts, and anyone intrigued by the ingenuity of secure communication in a bygone era. It will transport you to a time when the tap-tap-tap of Morse code carried the world's secrets, offering a captivating blend of historical context, technical ingenuity, and linguistic artistry

This annotated edition illustrates the capabilities of the AI Lab for Book-Lovers to add context and ease-of-use to manuscripts. It includes several types of abstracts, building from simplest to more complex: TLDR (one word), ELI5, TLDR (vanilla), Scientific Style, and Action Items; essays to increase viewpoint diversity, such as Grounds for Dissent, Red Team Critique, and MAGA Perspective; and Notable Passages and Nutshell Summaries for each page.

Annotations

TLDR (THREE WORDS):

Word dictionary.

ELI5

You have a super secret language only you and your friends know. This book is like a dictionary for that language. It has lots of short words that you can put together to make longer words. And it also has special instructions to help you understand the secret words, even if someone messes them up a little bit!

SCIENTIFIC STYLE ABSTRACT

John C. Hartfield's 'Lisbonian' Selection of Roots and Terminals," a telegraphic codebook published in 1911. Hartfield's codebook contains 21,323 five-letter codewords which can be combined to create 454,670,329 ten-letter codewords. The codewords are designed to be pronounceable and conform to the rules set by the Lisbon Telegraph Conference of 1908. The document includes a preface which describes the development of telegraphic codebooks and the rationale for Hartfield's system, instructions for using the codebook, a list of five-letter codewords divided into three sections (A, B, and C), a terminational order which lists possible two-letter endings for the codewords, and a detector which assists users in deciphering corrupted messages. The detectors are presented for both European and American Morse code.

OUTLINE

I. Introduction:

- **Title Page:** John C. Hartfield's "Lisbonian" Selection of Roots and Terminals
- **Preface:**
    - Explains the need for a standardized code language for telegraphic messages.
    - Highlights the abuse of the London Conference rules by code compilers.

    - Presents the amended rules of the Lisbon Telegraph Conference (1908) regarding artificial words.
    - Describes the compilation of 21,323 five-letter groups, equaling over 400 million ten-letter words.
    - Explains the use of Sections A and B for tabular work and Section C as roots.
    - Outlines the strengths of the selection:
        - Unique letter combinations
        - Telegraphic difference between letter groups
        - Minimized potential for errors due to spacing or signal conversion
        - Use of the Official Vocabulary and Standard Dictionaries
    - Includes testimonials from satisfied users (Dalgety & Company, Ltd.).
    - Mentions copyright protection and the use of detectors for error correction.

II. Instructions:

- Explains how to use the sections to create ten-letter words.
- Provides examples of how to form ten-letter words using the code.

III. Sections:

- **Section A:** Contains 20,000 five-letter groups, printed in black ink.
- **Section B:** Contains 20,000 five-letter groups, printed in brown ink.
- **Section C:** Contains 1,323 five-letter groups, used as roots.

IV. Terminational Order:

- This section presents a tabular arrangement of the 5-letter groups.
- The table helps users easily identify possible endings for ten-letter words.

V. Detectors:

- This section provides a guide to identify and correct errors in transmission.
- It includes tables for American and European Morse signals, as well as space letters.

VI. Signal Conversion Table:

- Helps users identify possible conversions of signals between American and European Morse.

VII. Sections A, B and C Index:

- Provides a quick index for navigating through the sections.

VIII. Appendix:

- May include additional information, such as:
    - A glossary of frequently used terms and phrases.
    - A table of numbers and their code equivalents.
    - Instructions for creating private codes using the "Lisbonian" selection.

Overall, the book aims to provide a comprehensive and reliable code system for telegraphic communication. Its key features include:

- **Extensive coverage:** Over 400 million unique ten-letter words.
- **Accuracy and reliability:** Careful construction to minimize errors.
- **Flexibility and ease of use:** Simple instructions and comprehensive terminational order.
- **Error correction tools:** Detectors and signal conversion table for efficient troubleshooting.

•

VIEWPOINTS

"LISBONIAN" AS A PRECURSOR TO "GENERATIVE AI"

Here's a comparison between Hartfield's "Lisbonian" effort and generative AI, emphasizing the similarities in regulating novel textual technologies:

Hartfield's "Lisbonian" Selection (Telegraph Codes)

- **Regulation:** The Lisbon Telegraph Conference (1908) aimed to regulate the use of artificial words, a new textual technology, to prevent ambiguity and security issues. They focused on:
  - **Pronounceability:** Ensuring words could be easily understood.
  - **Syllable structure:** Using recognizable syllables to reduce transmission errors.
  - **Length:** Limiting the number of letters per word to minimize errors.
- **Standardization:** The goal was to establish a common language for telegrams, promoting interoperability and clarity across different companies and countries.
- **Productization:** The selection was compiled and published as a commercial codebook, creating a market for this standardized technology.

Generative AI Today

- **Regulation:** The emerging field of generative AI faces similar regulatory challenges to telegraph codes in the early 20th century:
  - **Clarity and accuracy:** Ensuring generated text is understandable and accurate, avoiding misinformation.
  - **Structure and format:** Developing guidelines for AI-generated content, ensuring it conforms to acceptable formats and styles.

    - **Length and complexity:** Controlling the complexity and length of generated text, promoting ease of understanding and avoiding redundancy.
- **Standardization:** Efforts are underway to standardize generative AI, focusing on:
    - **Model transparency:** Establishing methods to understand how AI models work and make predictions.
    - **Data governance:** Developing best practices for training data, including bias mitigation and copyright compliance.
    - **Performance benchmarks:** Creating standardized ways to measure and compare AI model outputs.
- **Productization:** Generative AI models are being rapidly commercialized, with applications in content creation, marketing, customer service, and more.

Similarities in Regulation

- **Control over ambiguity and accuracy:** Both telegraph codes and generative AI require regulations to ensure clarity and prevent the spread of misleading or inaccurate information.
- **Emphasis on structure:** Both efforts emphasize the importance of structural rules in language to ensure clarity and minimize errors. Telegraph codes focused on pronounceable syllables, while generative AI requires guidelines for format, grammar, and style.
- **Preventing misuse:** Both face concerns about potential misuse, whether in the form of fraudulent telegrams or AI-generated disinformation campaigns.

Key Takeaway

The regulation of novel textual technologies like telegraph codes and generative AI share fundamental similarities. Both emphasize clarity, structure, and responsible use. The challenges of the digital age, however, require a more nuanced and sophisticated approach to regulation, considering broader ethical and societal implications.

Further Considerations

- **Speed of change:** Technological advancement in AI is far faster than in the telegraph era.
- **Scale and complexity:** Generative AI models are significantly more complex and capable than early code systems.
- **Global nature:** AI impacts a wider, more interconnected world, necessitating international collaboration on regulation.

Page-by-Page Summaries

BODY-1	*A selection of 21,323 five-letter words by John C. Hartfield, equivalent to 454,670,329 ten-letter words, with a clear and pronounceable character.*
BODY-2	*John C. Hartfield's "Lisbonian" selection of roots and terminal, containing 21,323 five-letter words equal to 454,670,329 ten-letter words, in accordance with the Lisbon Telegraph Conference of 1908.*
BODY-4	*The page discusses the need for a standardized code language for telegraphic messages, the abuse of fictitious pronounceable groups of letters, and the creation of 5-letter groups conforming to new rules to prevent errors and ensure security in telegraphic communication.*
BODY-5	*Testimonials for a code with over 36 million words used successfully for telegraph communication, with copyright information and examples of error correction using detectors provided.*
BODY-6	*Instructions for rearranging groups from Section A and Section B using figures 0, 1, 2, 3. Examples provided for memorization.*
BODY-49	*Section C of the document contains information related to a specific topic or category.*
BODY-52	*A list of numerical codes and corresponding words.*
BODY-53	*Termination order document with reference ID A101154.*
BODY-68	*Information about detectors.*
BODY-69	*A list of words and abbreviations organized alphabetically, with some missing letters and numbers.*
BODY-70	*The page contains a series of codes and abbreviations in a structured format, likely for reference or data analysis purposes.*
BODY-71	*The page contains a series of letters and abbreviations in a structured format, possibly related to a code or cipher.*
BODY-73	*The page contains a table of Morse code signals and their corresponding letters, along with instructions on how to detect errors in the signals and spacing.*
BODY-74	*Alphabet and Morse code symbols with corrections and changes.*
BODY-75	*Signal Conversion Table for quickly discovering signal conversions. Detector for space or "dot" letters, with American and European space letters listed for reference.*

Notable Passages

BODY-4	*"I have been experimenting on a selection of 5-Letter Groups, which will conform in every way to these Rules and at the same time be as safe under both European and American Morse as human ingenuity can make them, and in the construction of these 5 letter groups I have carefully checked them with the Official Vocabulary and have eliminated all 5 letter words, and all combinations of 2 and 3 letter words, which I found in that work, published at Berne under authority of the BUDA PESTH CONFERENCE and which still remains in force."*
BODY-5	*"Perhaps the following experience will serve best to illustrate the manner of using the Detectors to correct an error: When using my 36 Millions, I received the group 'zazon,' which could not be found. Referring to the Terminational Order, I found 'da' connected with 'zon,' but that did not make sense. In the same group was 'gu,' and on referring to the Detector, found 'gu' had exactly the same signals as 'za.' Working on that basis, it was found that 'guzon' gave the correct sense, which was verified later by the repetition of the word."*
BODY-75	*This Table will assist In discovering conversions of signals quickly.*

JOHN C. HARTFIELD'S

"LISBONIAN" SELECTION

OF

ROOTS AND TERMINALS,

CONTAINING

21,323 five letter words

EQUAL TO

454,670,329 ten letter words,

OF A CLEAR, EUPHONIOUS AND PRONOUNCEABLE CHARACTER.

JOHN C. HARTFIELD'S "LISBONIAN" SELECTION

OF

ROOTS AND TERMINALS,

CONTAINING

21,323 five letter words

EQUAL TO

454,670,329 ten letter words,

OF A CLEAR, EUPHONIOUS AND PRONOUNCEABLE CHARACTER.

With at least 2 letters difference between each group of 5 letters, and telegraphically dissimilar by

EUROPEAN AND AMERICAN MORSE.

Each syllable of which is in current use and can be verified by Dictionaries and the Official Vocabulary (*vide Preface*).

In strict accordance with the decision of the LISBON TELEGRAPH CONFERENCE, 1908, governing the use of Artificial Words.

No. 335

COMPILED BY

JOHN CHARLES HARTFIELD.

1911.

CABLES: "HARTFIELD, NEWYORK."

PRESS, HARTFIELD TELEGRAPHIC CODE PUBLISHING CO., NEW YORK.

EUROPEAN AGENT: EFFINGHAM WILSON, 54 Threadneedle St., LONDON, E. C.

TELEGRAMS: "EFFINGERE, LONDON."

TRADE MARK.

LEVIATHAN CODE

PREFACE.

For many years I have tried to obtain the consent of the Governments and Cable Companies to the recognition of one Code language for Telegraphic messages. It was through my constant agitation, backed by a Petition signed by the leading cablers of the world, for the permissive use of "HARTFIELD'S ROOTS and TERMINALS," that the LONDON CONFERENCE passed a rule admitting the use of fictitious pronounceable groups of letters (limited to 10). This rule has been greatly abused by the publication of Thousands of Millions of groups of letters, many differing only one letter or only one telegraphic signal and quite unpronounceable. This has not only caused the Governments and Cable Companies much loss and inconvenience, on account of the numerous repetitions demanded by their customers, but has been most unfair to those Cablers and Code Compilers who have honestly adhered to the Rule in the construction of their Codes. This abuse was severely commented upon by the British Postmaster (vide Pamphlet, p. 10).

The LISBON CONFERENCE, in order to place a check on illegal combinations of letters and still give the cabling public the great advantage of fictitious words properly constructed, amended the LONDON CONFERENCE Rules as follows:

2. The words, whether genuine or artificial, must be formed of syllables capable of pronunciation according to the **current usage** of one of the following languages:—German, English, Spanish, French, Dutch, Italian, Portuguese, or Latin.
3. Words in code language must not be longer than ten characters according to the Morse alphabet, the combinations ae, aa, ao, oe, ue, being counted as two letters each. The combination ch is also counted as two letters in artificial words.
4. Combinations not fulfilling the conditions of the two preceding paragraphs are regarded as language in letters having a secret meaning and are charged for accordingly. Nevertheless, those which may be formed by the union of two or more words in plain language contrary to the usage of the language are not admitted.

I have been experimenting on a selection of 5-Letter Groups, which will conform in every way to these Rules and at the same time be as safe under both European and American Morse as human ingenuity can make them, and in the construction of these 5 letter groups I have carefully checked them with the **Official Vocabulary** and have eliminated all 5 letter words, and all combinations of 2 and 3 letter and 3 and 2 letter words, which I found in that work, published at Berne under authority of the BUDA PESTH CONFERENCE and which still remains in force.

I have now completed 21,323 groups of 5 letters, giving a range of 454,670,329 Ten Letter Words. I beg to suggest that Section "A" and Section "B," containing 20,000 5 letter groups, be employed for Tabular work, and Section "C," containing 1,323 groups, be used **only** as **Roots**, connecting with the 20,000 groups in Sections "A" and "B," the latter being used as **Terminals.** One thus obtains 1,323 times 20,000, or over 26 Millions of **distinctive** ten letter words for phrases only.

I make the following claims for these words:

(*a*) Each group of five Letters has at least a difference of two letters from any other group, and transpositions in the last two letters (which have frequently caused trouble) have been carefully avoided. Take the group commencing "abd" as an example. I have **abdaf, abded, abdil, abdoc, abduz, abdyx.** On examination of the two final letters of these groups, a repetition of a letter will not be found. In the construction of the 1st and 2nd letters connecting with the same 3rd, 4th and 5th letters, I have made as wide a telegraphic difference as possible, and here also avoided transpositions, as will be seen by the following group, **ab, bi, co, da, ev, gu, ij, le, my, on, uk, ys,** connecting with abo, acs, &c.

(*b*) I have made a telegraphic difference between each group of two letters to avoid errors that occur from wrong spacing of signals; for instance, European Morse "**guzon**" and "**zazon**" have the same signals "g — — . u . . —" "z — — . . a . —" and by American Morse "**esbub**" and "**evcub**" have the same signals "s . . . b — . . ." "v . . . — c . . ."

(*c*) I have avoided as much as possible the chances of two errors of wrong spacing, first by European Morse and second by American Morse; for instance, "**fadus**"(a . — d — . .), by an error between Calcutta and New York, could become "**fezus**" (e . z — — . .) and the final s . . . (by a slight error between New York and Chicago) become r . . . So the word "**fadus**" as it left Calcutta might be delivered in Chicago "**fezur**."

In the event of dislocation of traffic from the East to Europe, causing the Cable Companies to use the American routes, this avoidance of similarities is of great importance to European, Asiatic and Australasian Cablers, but of paramount importance to American telegraphic correspondents with Asia and Australasia even under normal conditions.

(*d*) I have taken the Official Vocabulary and Standard Dictionaries as my guide and can guarantee that every group of 5 letters not only consists of syllables **in current use** in the eight languages, but they can be verified by the **Official Vocabulary.**

The search for euphonious syllables has been a long, tedious and trying piece of work. Many a one has been found only after much research. Take for instance "**ryb**" found in Eu"ryb"ates; "vus" found in Cor"vus"; "ams" found in **Brough**"ams"; "yb" found in Mol"yb"dena.

For obvious reasons, I do not publish my key to the syllables I employ, but I am prepared at any time to give my authority for any syllable that may be questioned.

In view of the fact that the total number of syllables employed in this selection is only about 2,700, a greater security from error is obtained than can be found in many Selections of less elasticity. I have worked on the same plan as I did when I made the 36 Millions (10 letter words) first employed in the Private Code of **Messrs. Dalgety & Co., Ltd., Australian Pastoral Merchants.** I can honestly say that I believe this Selection of

PREFACE (continued).

over 400 Millions (10 letter words) will be found to work quite as satisfactorily as the 36 Millions. **Messrs. Dalgety & Co., Ltd.,** passed between **London** and **Sydney** in one day **nearly 600 words without an error,** and when one considers that they have over 50 Branch Houses in Australasia in hourly communication [the following testimonials will be appreciated:

COPY OF TESTIMONIALS.

96 Bishopsgate Street, Within,
London, E. C.,
17th September, 1909.

John C. Hartfield, Esq.,
Dear Sir:

We have used your 36 Millions of words between our numerous branches for over 3 years, during which time a large number of words have been telegraphed daily. The mutilations in transmission have been comparatively few, and those that have occurred, have, in almost every case, been capable of correction by the Detectors.

Yours truly,
For DALGETY & COMPANY, Limited,
(Signed) Thos. B. Fisher, Secretary.

Cable from JOHN C. HARTFIELD,
New York.

To Messrs. Dalgety & Company, Ltd.
London,

July 12th, 1911.

May I publish that your Code continues as usual working well.

Cable from DALGETY & COMPANY, Ltd.,
London.
To John C. Hartfield.
New York.

July 13th, 1911.

Referring to your Cable of July 12th, the reply is "Yes."

COPYRIGHT.

The attention of would-be copyists of any portion of this work is called to the following extracts from the **Copyright Law,** [Sec. 3 and Sec. 6:

Sec. 3. **That the copyright provided by this Act shall protect all the copyrightable component parts of the work copyrighted,** and all matter therein in which copyright is already subsisting, but without extending the duration or scope of such copyright. **The copyright upon composite works or periodicals shall give to the proprietor thereof all the rights in respect thereto which he would have if each part were individually [copyrighted under this Act.**

Sec. 6. That compilations or abridgments, adaptations, arrangements, dramatizations, translations, or other versions of works in the public domain, or of copyrighted works when produced with the consent of the proprietor of the copyright of such works, or works republished with new matter, shall be regarded as new works subject to copyright under the provisions of this Act; but the publication of any such new works shall not affect the force or validity of any subsisting copyright upon the matter employed or any part thereof, or be construed to imply an exclusive right to such use of the original works, or to secure or extend copyright in [such original works.

Estimates will be furnished on application for compiling and/or printing Private Codes, including the use of this "**Lisbonian**" selection, but the right of reproduction of any [portion is strictly reserved according to law.

USE OF THE DETECTORS.

Perhaps the following experience will serve best to illustrate the manner of using the Detectors to correct an error: When using my 36 Millions, I received the group "zazon," which could not be found. Referring to the Terminational Order, I found "da" connected with "zon," but that did not make sense. In the same group was "gu," and on referring to the Detector, found "gu" had exactly the same signals as "za." Working on that basis, it was found that "guzon" gave the correct sense, which was [verified later by the repetition of the word.

The letters and figures in heavy type refer to American Morse only; for instance: In European Morse "**ea**" (. . —) are equivalent to "**1t**" "**u,**" but in American Morse those 2 letters are equivalent to "**2 1**" "**2 t,**" *i. e.* "**11**" "**o1**" "**1t**" "**ot.**" The figures represent the number of Dots as per Chart of Dot (space) letters.

JOHN C. HARTFIELD
(Leviathan Code).

INSTRUCTIONS.

For First figure 0 take First and Second groups from Section A
" " " 1 " First group from Section A and Second group from (Section B
" " " 2 " First and Second groups from Section B
" " " 3 " First group from Section B and Second group from (Section A

or

A	**A**	=	**First figure**	**0**	=	**A**	**A**
A	**B**	=	**First figure**	**1**	=	**A**	**B**
B	**B**	=	**First figure**	**2**	=	**B**	**B**
B	**A**	=	**First figure**	**3**	=	**B**	**A**

The remaining figures becoming 2nd, 3rd, 4th, 5th, 6th, 7th, 8th and 9th.

EXAMPLES:

029647326	cunavhysif
164784295	gezalpofut
253467010	seyomukdaf
349523000	rovipcyaud

cunavhysif	A	2964	A	7326	=	029647326
gezalpofut	A	6478	B	4295	=	164784295
seyomukdaf	B	5346	B	7010	=	253467010
rovipcyaud	B	4952	A	3000	=	349523000

To assist in memorizing the Sections, "A" is printed in Black Ink and "B" in Brown Ink.

SECTION A.

SECTION A

0000 ababo	**0050** abmyl	**0100** acdon	**0150** acoda	**0200** aderl	**0250** afabu	**0300** afkol	**0350** afvan	**0400** agjuk	**0450** agvel
0001 abacs	0051 abnes	0101 acdry	0151 acofi	0201 adesy	0251 afads	0301 afkub	0351 afvik	0401 agkac	0451 agvij
0002 abald	0052 abnic	0102 acduk	0152 acoku	0202 adewa	0252 afafi	0302 aflaf	0352 afvum	0402 agkeg	0452 agvot
0003 abapy	0053 abobu	0103 acebu	0153 acowl	0203 adexe	0253 afalm	0303 afled	0353 afwar	0403 agkit	0453 agwan
0004 abaxi	0054 abogy	0104 acele	0154 acpam	0204 adfab	0254 afane	0304 aflip	0354 afwes	0404 agkul	0454 agyat
0005 abaza	0055 abols	0105 acens	0155 acpel	0205 adfev	0255 afaty	0305 afloc	0355 afwot	0405 agkyn	0455 agzeh
0006 abbif	0056 abond	0106 aceph	0156 acpiv	0206 adfip	0256 afawk	0306 afluz	0356 afyaw	0406 aglad	0456 agzof
0007 abblu	0057 aboze	0107 acerd	0157 acpot	0207 adflu	0257 afbas	0307 aflyn	0357 afzey	0407 aglec	0457 ahaep
0008 abbro	0058 abpar	0108 acewi	0158 acpud	0208 adfon	0258 afbek	0308 afmag	0358 afzog	0408 agliv	0458 ahait
0009 abcec	0059 abpez	0109 acfac	0159 acran	0209 adgac	0259 afbli	0309 afmeb	0359 afzul	0409 aglol	0459 ahand
0010 abcyn	**0060** abpik	**0110** acfeg	**0160** acrew	**0210** adgeg	**0260** afbor	**0310** afmil	**0360** aganz	**0410** aglub	**0460** ahary
0011 abdaf	0061 abpog	0111 acfib	0161 acrod	0211 adgin	0261 afbug	0311 afmox	0361 agask	0411 agmaf	0461 ahbat
0012 abded	0062 abpyt	0112 acfom	0162 acrum	0212 adgom	0262 afcav	0312 afmun	0362 agath	0412 agmed	0462 ahbeb
0013 abdil	0063 abras	0113 acfru	0163 acsap	0213 adgyl	0263 afcit	0313 afmyc	0363 agauc	0413 agmoc	0463 ahbir
0014 abdoc	0064 abrek	0114 acgad	0164 acses	0214 adhad	0264 afcop	0314 afnah	0364 agaxo	0414 agmuz	0464 ahbow
0015 abduz	0065 abror	0115 acgec	0165 acsof	0215 adhif	0265 afcur	0315 afnef	0365 agbar	0415 agnag	0465 ahbul
0016 abdyx	0066 absav	0116 acgol	0166 acsuc	0216 adhyr	0266 afcyg	0316 afnib	0366 agbez	0416 agneb	0466 ahcal
0017 abeba	0067 abtre	0117 acgub	0167 acsyr	0217 adinz	0267 afdax	0317 afnoz	0367 agbim	0417 agnis	0467 ahcef
0018 abech	0068 abube	0118 achav	0168 actaw	0218 adiro	0268 afder	0318 afnyl	0368 agbly	0418 agnox	0468 ahoot
0019 abege	0069 abuja	0119 achoy	0169 actex	0219 adjeh	0269 afdiz	0319 afoba	0369 agbog	0419 agnun	0469 ahdad
0020 abeki	**0070** abult	**0120** achul	**0170** actuy	**0220** adkaf	**0270** afdov	**0320** afoky	**0370** agbux	**0420** agobi	**0470** ahdec
0021 abemp	0071 abumy	0121 acica	0171 acuga	0221 adked	0271 afdup	0321 aford	0371 agcas	0421 agodu	0471 ahdol
0022 abesu	0072 abunz	0122 acifo	0172 acvar	0222 adkli	0272 afemi	0322 afote	0372 agcin	0422 agoft	0472 ahdub
0023 abfeb	0073 abvov	0123 acigu	0173 acvez	0223 adkoc	0273 afepa	0323 afown	0373 agcor	0423 agogo	0473 ahdym
0024 abgat	0074 abvup	0124 acimp	0174 acvin	0224 adliz	0274 afest	0324 afoxo	0374 agcug	0424 agoha	0474 ahemb
0025 abgef	0075 abwaz	0125 acits	0175 acvog	0225 adlox	0275 afewe	0325 afpak	0375 agcyb	0425 agove	0475 aheps
0026 abgid	0076 abwod	0126 acjaf	0176 acwat	0226 adlun	0276 affaz	0326 afpem	0376 agdav	0426 agoys	0476 aherz
0027 abgoz	0077 abyus	0127 acjed	0177 acwid	0227 adlyt	0277 affep	0327 afpid	0377 agdej	0427 agpaw	0477 ahfed
0028 abgul	0078 abzah	0128 acjoc	0178 acyah	0228 admef	0278 affly	0328 afplo	0378 agdib	0428 agpir	0478 ahfis
0029 abgyb	0079 abzin	0129 acjug	0179 acyen	0229 admic	0279 affok	0329 afpry	0379 agdop	0429 agpoz	0479 ahgaw
0030 abhok	**0080** acaba	**0130** ackeb	**0180** acyow	**0230** admoz	**0280** affus	**0330** afpuf	**0380** agebs	**0430** agput	**0480** ahget
0031 abhyc	0081 acach	0131 ackil	0181 aczet	0231 adnem	0281 afgab	0331 afral	0381 agegi	0431 agrak	0481 ahgig
0032 abibi	0082 acaim	0132 ackun	0182 aczop	0232 adnid	0282 afgis	0332 afren	0382 agepe	0432 agrem	0482 ahgro
0033 abicu	0083 acaju	0133 ackyd	0183 aczur	0233 adofu	0283 afgon	0333 afrij	0383 agezu	0433 agrig	0483 ahguy
0034 abiko	0084 acaly	0134 aclay	0184 adabi	0234 adoly	0284 afgyr	0334 afrut	0384 agfax	0434 agrob	0484 ahhan
0035 abing	0085 acant	0135 aclef	0185 adady	0235 adort	0285 afhap	0335 afsam	0385 agfer	0435 agruf	0485 ahhyl
0036 abisy	0086 acase	0136 aclig	0186 adapo	0236 adpos	0286 afhew	0336 afsel	0386 agfid	0436 agsal	0486 ahims
0037 abith	0087 acazo	0137 acloz	0187 adaux	0237 adrez	0287 afhof	0337 afsho	0387 aggay	0437 agscu	0487 ahion
0038 abjak	0088 acbax	0138 acluv	0188 adbav	0238 adrhi	0288 afhuc	0338 afsif	0388 aggep	0438 agsen	0488 ahirp
0039 abjem	0089 acber	0139 acmak	0189 adbij	0239 adryn	0289 afhyd	0339 afsud	0389 aggil	0439 agski	0489 ahitz
0040 abjuf	**0090** acblo	**0140** acmem	**0190** adboh	**0240** adtas	**0290** afibs	**0340** afsyk	**0390** aghaz	**0440** agsos	**0490** ahjav
0041 abkal	0091 acbup	0141 acmip	0191 adcax	0241 adtey	0291 afiod	0341 aftat	0391 aghex	0441 agtap	0491 ahjib
0042 abkyk	0092 acbyn	0142 acmob	0192 adcer	0242 adtik	0292 afipi	0342 aftez	0392 aghik	0442 agtew	0492 ahjod
0043 ablam	0093 accaz	0143 acmuf	0193 adcib	0243 adunc	0293 afirl	0343 aftig	0393 aghum	0443 agtiz	0493 ahjum
0044 ablel	0094 accep	0144 acmyg	0194 adcov	0244 adurg	0294 afjac	0344 aftoy	0394 aghyp	0444 agtod	0494 ahkay
0045 ablix	0095 accro	0145 acnal	0195 addaz	0245 aduze	0295 afjeg	0345 aftru	0395 agimu	0445 agtur	0495 ahkex
0046 ablud	0096 accus	0146 acnek	0196 addep	0246 adves	0296 afjuw	0346 afuma	0396 agipo	0446 agtyl	0496 ahklo
0047 abmex	0097 acdab	0147 acnix	0197 addok	0247 adviv	0297 afkad	0347 afund	0397 agjab	0447 aguca	0497 ahkut
0048 abmig	0098 acdev	0148 acnyo	0198 adedi	0248 adwor	0298 afkec	0348 afuse	0398 agjif	0448 agure	0498 ahlab
0049 abmoh	0099 acdic	0149 acobs	0199 adeng	0249 adxel	0299 afkir	0349 afuti	0399 agjon	0449 agvam	0499 ahley

0500 ahlil	**0550** ajcic	**0600** ajned	**0650** akces	**0700** akmif	**0750** alboy	**0800** alraf	**0850** amebe	**0900** amorb	**0950** anefi
0501 ahlop	0551 ajcog	0601 ajnoc	0651 akciv	0701 akmom	0751 albud	0801 alred	0851 amecs	0901 amoxa	0951 anega
0502 ahluk	0552 ajcux	0602 ajnuz	0652 akcof	0702 aknew	0752 alcac	0802 alril	0852 amerf	0902 ampeg	0952 aneke
0503 ahmew	0553 ajdas	0603 ajodi	0653 akcuc	0703 aknik	0753 alceg	0803 alroc	0853 ameta	0903 ampin	0953 anelm
0504 ahmoy	0554 ajdek	0604 ajoho	0654 akcyx	0704 aknol	0754 alcom	0804 alruz	0854 amewo	0904 amplu	0954 anepu
0505 ahnin	0555 ajdor	0605 ajonu	0655 akdar	0705 akobo	0755 alcys	0805 alsag	0855 amfan	0905 ampom	0955 anery
0506 ahnos	0556 ajdug	0606 ajost	0656 akdez	0706 akofy	0756 aldan	0806 alseb	0856 amfew	0906 amrad	0956 anfam
0507 ahnyt	0557 ajdyc	0607 ajoxy	0657 akdip	0707 akome	0757 aldey	0807 alsik	0857 amfod	0907 amrec	0957 anfel
0508 ahoaf	0558 ajeid	0608 ajpag	0658 akdog	0708 akonc	0758 aldif	0808 alsox	0858 amfri	0908 amrol	0958 anfig
0509 ahoen	0559 ajelz	0609 ajpeb	0659 akdux	0709 akopa	0759 aldod	0809 alsun	0859 amfum	0909 amrub	0959 anfot
0510 ahora	**0560** ajemy	**0610** ajpim	**0660** akeat	**0710** akori	**0760** aldut	**0810** alsyd	**0860** amfys	**0910** amsaf	**0960** anfud
0511 ahoux	0561 ajept	0611 ajpow	0661 akedy	0711 akots	0761 alecu	0811 altah	0861 amgap	0911 amsed	0961 angan
0512 ahows	0562 ajere	0612 ajpun	0662 akefo	0712 akpaf	0762 alego	0812 altob	0862 amges	0912 amshy	0962 angic
0513 ahozo	0563 ajevu	0613 ajraw	0663 akelv	0713 akped	0763 alerx	0813 altyn	0863 amgiz	0913 amsoc	0963 angod
0514 ahpip	0564 ajfav	0614 ajref	0664 akemu	0714 akpoc	0764 alews	0814 aluky	0864 amgof	0914 amsuz	0964 angum
0515 ahpla	0565 ajfey	0615 ajroz	0665 akerg	0715 akpuz	0765 alfap	0815 alume	0865 amguc	0915 amtav	0965 anhay
0516 ahpof	0566 ajfiz	0616 ajsah	0666 akfas	0716 akrag	0766 alfit	0816 aluph	0866 amgym	0916 amtem	0966 anhet
0517 ahpre	0567 ajfop	0617 ajsem	0667 akfix	0717 akreb	0767 alfro	0817 aluvi	0867 amhey	0917 amtis	0967 anhor
0518 ahpuh	0568 ajfur	0618 ajsip	0668 akfor	0718 akrid	0768 alful	0818 aluzu	0868 amhip	0918 amtuf	0968 anhun
0519 ahrog	0569 ajgax	0619 ajsob	0669 akfug	0719 akrun	0769 algar	0819 alvay	0869 amhoz	0919 amubs	0969 anhym
0520 ahrys	**0570** ajger	**0620** ajsuf	**0670** akgav	**0720** akryf	**0770** algog	**0820** alvef	**0870** amhux	**0920** amuly	**0970** aniei
0521 ahsak	0571 ajgib	0621 ajtam	0671 akgim	0721 aksaw	0771 alhat	0821 alvid	0871 amich	0921 amuno	0971 anigy
0522 ahsme	0572 ajgov	0622 ajtep	0672 akgop	0722 aksef	0772 alhib	0822 alvoz	0872 amine	0922 amuzi	0972 aniku
0523 ahtag	0573 ajgup	0623 ajtiv	0673 akgur	0723 aksit	0773 alhov	0823 alwaw	0873 amiza	0923 amvag	0973 anilo
0524 ahtek	0574 ajhir	0624 ajtot	0674 akgyn	0724 aksoz	0774 alhuf	0824 alwem	0874 amjar	0924 amvil	0974 anjap
0525 ahtho	0575 ajhox	0625 ajtul	0675 akhab	0725 aksuv	0775 alhyg	0825 alwis	0875 amjez	0925 amvox	0975 anjes
0526 ahtuz	0576 ajhuk	0626 ajudy	0676 akhek	0726 aksys	0776 aliba	0826 alyak	0876 amjog	0926 amvun	0976 anjil
0527 ahupu	0577 ajhys	0627 ajval	0677 akhup	0727 aktha	0777 aliru	0827 alzal	0877 amkas	0927 amwir	0977 anjof
0528 ahvur	0578 ajics	0628 ajven	0678 akimo	0728 aktry	0778 aljas	0828 alzen	0878 amkek	0928 amwob	0978 anjux
0529 ahwam	0579 ajigo	0629 ajvit	0679 akiny	0729 aktuk	0779 aljet	0829 alzip	0879 amkid	0929 amyup	0979 ankar
0530 ahxem	**0580** ajika	**0630** ajvos	**0680** akipu	**0730** akufe	**0780** alkav	**0830** alzos	**0880** amkor	**0930** amzic	**0980** ankib
0531 ahxyn	0581 ajiph	0631 ajway	0681 akiti	0731 akuku	0781 alkry	0831 alzug	0881 amkug	0931 amzuk	0981 ankog
0532 ahyap	0582 ajity	0632 ajwex	0682 akiwa	0732 akuni	0782 almaz	0832 amaco	0882 amkyp	0932 anacy	0982 anlas
0533 ahzor	0583 ajizu	0633 akabe	0683 akjah	0733 akvak	0783 almep	0833 amada	0883 amlaw	0933 anadu	0983 anlek
0534 ahzuc	0584 ajkab	0634 akack	0684 akjer	0734 akvem	0784 almin	0834 amaik	0884 amleh	0934 anale	0984 anlow
0535 ajahs	0585 ajkev	0635 akaft	0685 akjis	0735 akvic	0785 almok	0835 amarg	0885 amlit	0935 anani	0985 anlug
0536 ajaki	0586 ajkis	0636 akagi	0686 akkaz	0736 akvul	0786 alnab	0836 amavu	0886 amlyd	0936 anarx	0986 anlyb
0537 ajalb	0587 ajklu	0637 akahu	0687 akkep	0737 akweh	0787 alnev	0837 ambal	0887 ammax	0937 anbak	0987 anmeh
0538 ajamt	0588 ajkon	0638 akaja	0688 akkin	0738 akyle	0788 alnig	0838 amben	0888 ammib	0938 anbem	0988 anmik
0539 ajaud	0589 ajkyr	0639 akamy	0689 akkok	0739 akyot	0789 alnon	0839 ambix	0889 ammov	0939 anbob	0989 anmop
0540 ajawe	**0590** ajlac	**0640** akarn	**0690** akkus	**0740** akyub	**0790** alnuk	**0840** amcab	**0890** ammut	**0940** anbuf	**0990** anmur
0541 ajaya	0591 ajleg	0641 akasp	0691 akkym	0741 akzam	0791 alnyp	0841 amcli	0891 ammyn	0941 anbyx	0991 anmyd
0542 ajbap	0592 ajlif	0642 akawo	0692 aklal	0742 akzel	0792 alody	0842 amcon	0892 amnaz	0942 ancaw	0992 annax
0543 ajbes	0593 ajlom	0643 akban	0693 aklex	0743 akzoy	0793 alola	0843 amcre	0893 amnep	0943 ancoy	0993 anner
0544 ajbig	0594 ajlyp	0644 akbet	0694 aklib	0744 alaci	0794 alour	0844 amdam	0894 amnim	0944 ancut	0994 annif
0545 ajbof	0595 ajmad	0645 akbiz	0695 akloh	0745 alafu	0795 alpec	0845 amdel	0895 amnok	0945 andac	0995 annov
0546 ajbuc	0596 ajmec	0646 akbod	0696 aklut	0746 alalk	0796 alpij	0846 amdiv	0896 amnus	0946 andeg	0996 annup
0547 ajbyb	0597 ajmix	0647 akbum	0697 aklyr	0747 alash	0797 alpol	0847 amdro	0897 amofo	0947 andin	0997 annyg
0548 ajcar	0598 ajmol	0648 akbyc	0698 akmac	0748 albel	0798 alpub	0848 amdud	0898 amoke	0948 andul	0998 anoch
0549 ajchy	0599 ajnaf	0649 akcap	0699 akmeg	0749 albiv	0799 alpyr	0849 amdyt	0899 amopu	0949 andyp	0999 anomi

anops anorf anpaz anpep

[redacted]

SECTION A

1000 anpok	**1050** apdeb	**1100** apniv	**1150** aqdre	**1200** ardak	**1250** arpev	**1300** asegu	**1350** aspur	**1400** atgoc	**1450** atude
1001 anpus	1051 apdid	1101 apnom	1151 aqelo	1201 ardem	1251 arpif	1301 aseko	1351 asrif	1401 atgre	1451 atufi
1002 anrab	1052 apdox	1102 apnys	1152 aqerb	1202 ardig	1252 arpuk	1302 asems	1352 asrok	1402 atgud	1452 atuor
1003 anrev	1053 apdun	1103 apoce	1153 aqetz	1203 ardob	1253 arpyc	1303 asepy	1353 asrud	1403 athac	1453 atvis
1004 anriz	1054 apeky	1104 apoli	1154 aqfuc	1204 arduf	1254 arrax	1304 asexa	1354 asryc	1404 athez	1454 atwaf
1005 anron	1055 apeld	1105 apoms	1155 aqgah	1205 ardyn	1255 arrov	1305 asezi	1355 assev	1405 athid	1455 atwel
1006 anruk	1056 apepi	1106 aponz	1156 aqhef	1206 arebo	1256 arrup	1306 asfah	1356 asson	1406 atini	1456 atwog
1007 ansad	1057 aperu	1107 apoth	1157 aqhim	1207 arede	1257 arryd	1307 asfem	1357 assym	1407 atisk	1457 atxar
1008 ansec	1058 apete	1108 apowa	1158 aqhos	1208 areku	1258 arsow	1308 asfyl	1358 astad	1408 atjad	1458 atxen
1009 ansol	1059 apfat	1109 apoxu	1159 aqial	1209 arela	1259 arsty	1309 asgal	1359 astec	1409 atjek	1459 atxim
1010 anspy	**1060** apfeh	**1110** appax	**1160** aqkom	**1210** arern	**1260** arteg	**1310** asgen	**1360** astub	**1410** atjow	**1460** atyas
1011 ansti	1061 apfim	1111 apper	1161 aqmym	1211 areti	1261 artin	1311 asgiv	1361 astyp	1411 atjur	1461 atyef
1012 ansub	1062 apfoz	1112 appic	1162 aqnat	1212 arfaw	1262 artum	1312 asgos	1362 asubi	1412 atkaw	1462 atygo
1013 anted	1063 apgak	1113 appov	1163 aqofe	1213 arfoy	1263 artyx	1313 asgut	1363 asuge	1413 atkem	1463 atyne
1014 antip	1064 apgem	1114 appup	1164 aqohn	1214 arfut	1264 arucu	1314 asheh	1364 asulo	1414 atkot	1464 atyum
1015 antuc	1065 apgly	1115 aprul	1165 aqois	1215 argam	1265 arung	1315 ashin	1365 asups	1415 atkuk	1465 atzap
1016 antyt	1066 apgob	1116 aprym	1166 aqoty	1216 argel	1266 arupa	1316 ashuy	1366 asuxu	1416 atlav	1466 atzif
1017 anubo	1067 apguf	1117 apsar	1167 aqoud	1217 argip	1267 aruso	1317 asidu	1367 asuza	1417 atlep	1467 atzus
1018 anuka	1068 aphil	1118 apsez	1168 aqpri	1218 argru	1268 arvad	1318 asily	1368 asvac	1418 atlir	1468 avacu
1019 anulu	1069 aphow	1119 apsog	1169 aqret	1219 arhaf	1269 arvec	1319 asize	1369 asveg	1419 atlok	1469 avams
1020 anune	**1070** apibu	**1120** apsux	**1170** aqrin	**1220** arher	**1270** arvir	**1320** asjel	**1370** asvom	**1420** atlyc	**1470** avata
1021 anvat	1071 apify	1121 aptaz	1171 aqroy	1221 arhuz	1271 arvol	1321 asjig	1371 aswak	1421 atmat	1471 avave
1022 anvix	1072 apige	1122 aptix	1172 aqsca	1222 arifu	1272 arwac	1322 asjus	1372 aswic	1422 atmey	1472 avaxy
1023 anvoc	1073 apind	1123 aptyr	1173 aqshe	1223 arilk	1273 arwed	1323 askan	1373 aswol	1423 atmod	1473 avazi
1024 anwal	1074 apirk	1124 apudi	1174 aqsiv	1224 arivy	1274 aryls	1324 askey	1374 aszax	1424 atmup	1474 avbab
1025 anwiv	1075 apisl	1125 apugu	1175 aqsny	1225 arjan	1275 aryoc	1325 askip	1375 aszer	1425 atnak	1475 avbev
1026 anxis	1076 apivo	1126 apuko	1176 aqsok	1226 arjeb	1276 arzat	1326 askod	1376 ataam	1426 atnec	1476 avbis
1027 anyeb	1077 apixa	1127 apule	1177 aqsut	1227 arjik	1277 asahy	1327 askum	1377 atabs	1427 atniz	1477 avbom
1028 anyom	1078 apjal	1128 apuna	1178 aqtay	1228 arkap	1278 asaix	1328 aslap	1378 atake	1428 atnof	1478 avbuk
1029 anzag	1079 apjen	1129 apvas	1179 aqtle	1229 arkes	1279 asang	1329 asles	1379 atalu	1429 atnut	1479 avcem
1030 anzoz	**1080** apjos	**1130** apvej	**1180** aqtro	**1230** arkuc	**1280** asape	**1330** aslik	**1380** atarc	**1430** atoby	**1480** avciz
1031 apagy	1081 apjut	1131 apvib	1181 aquha	1231 arlar	1281 asara	1331 aslof	1381 atavi	1431 atocs	1481 avcob
1032 apala	1082 apkam	1132 apvor	1182 aqury	1232 arlog	1282 asaul	1332 asluc	1382 atbex	1432 atoka	1482 avcra
1033 apano	1083 apkel	1133 apwah	1183 aqvop	1233 arlux	1283 asbaf	1333 asmez	1383 atcah	1433 atotz	1483 avcyc
1034 apare	1084 apkif	1134 apwey	1184 aqwas	1234 arlym	1284 asbed	1334 asmid	1384 atcig	1434 atoxi	1484 avdap
1035 apasi	1085 apkro	1135 apwho	1185 aqwep	1235 armas	1285 asbit	1335 asmog	1385 atcul	1435 atpaj	1485 avdof
1036 apatu	1086 apkud	1136 apyon	1186 aqwib	1236 armek	1286 asboc	1336 asmux	1386 atcym	1436 atpho	1486 avduc
1037 apawl	1087 aplan	1137 apyra	1187 aradz	1237 armit	1287 asbry	1337 asnas	1387 atdet	1437 atray	1487 aveha
1038 apbad	1088 aplet	1138 apyug	1188 aralc	1238 armug	1288 asbuz	1338 asnor	1388 atdom	1438 atroh	1488 avelu
1039 apbec	1089 aplis	1139 apzab	1189 arari	1239 arnav	1289 ascag	1339 asnug	1389 atdri	1439 atsib	1489 aveny
1040 apbik	**1090** aplod	**1140** apzef	**1190** arasm	**1240** arneh	**1290** asceb	**1340** asnyx	**1390** atdyl	**1440** atsko	**1490** avepo
1041 apboi	1091 aplum	1141 apzit	1191 araxu	1241 arnop	1291 ascil	1341 asoga	1391 atebi	1441 atsla	1491 avets
1042 apbub	1092 apmap	1142 aqaig	1192 arbag	1242 arnur	1292 ascox	1342 asolu	1392 ateft	1442 atstu	1492 avexi
1043 apcaf	1093 apmes	1143 aqalt	1193 arbef	1243 arock	1293 ascun	1343 asopi	1393 atelk	1443 atswe	1493 aveze
1044 apced	1094 apmiz	1144 aqaon	1194 arbox	1244 arolo	1294 asday	1344 asovo	1394 aterp	1444 atsyf	1494 avfay
1045 apchi	1095 apmof	1145 aqaum	1195 arbun	1245 aromu	1295 asdef	1345 asoxe	1395 atesh	1445 attan	1495 avfex
1046 apcoc	1096 apmuc	1146 aqbra	1196 arbyl	1246 aroni	1296 asdim	1346 aspav	1396 ateuc	1446 attop	1496 avfil
1047 apcuz	1097 apmyt	1147 aqcla	1197 arcay	1247 aropy	1297 asdoz	1347 aspex	1397 atfol	1447 attux	1497 avfoh
1048 apcyk	1098 apnac	1148 aqcyt	1198 arcet	1248 arota	1298 asduv	1348 aspli	1398 atfra	1448 atual	1498 avfuz
1049 apdag	1099 apneg	1149 aqdow	1199 arcoz	1249 arpab	1299 asect	1349 aspop	1399 atgli	1449 atubu	1499 avgas

1500 avgek	**1550** avtax	**1600** awfli	**1650** awtug	**1700** axjup	**1750** ayaeg	**1800** ayrel	**1850** azexu	**1900** azsac	**1950** baele
1501 avgit	1551 avter	1601 awfos	1651 awtyb	1701 axkak	1751 ayaka	1801 ayroj	1851 azfar	1901 azseg	1951 baens
1502 avgor	1552 avtup	1602 awfre	1652 awufy	1702 axkeh	1752 ayamp	1802 ayruw	1852 azfez	1902 azsir	1952 baeph
1503 avgug	1553 avuby	1603 awglu	1653 awumo	1703 axkob	1753 ayanu	1803 ayske	1853 azfik	1903 azsom	1953 baerd
1504 avheo	1554 avufu	1604 awham	1654 awupi	1704 axkys	1754 ayard	1804 aysov	1854 azfog	1904 azsyn	1954 baeto
1505 avhiv	1555 avujo	1605 awheb	1655 awuvu	1705 axlov	1755 ayasy	1805 ayspu	1855 azfux	1905 aztak	1955 baewi
1506 avids	1556 avulk	1606 awhud	1656 awvep	1706 axlul	1756 ayaxt	1806 aytaf	1856 azgaf	1906 aztet	1956 bafac
1507 aviga	1557 avvaz	1607 awiap	1657 awvod	1707 axlyg	1757 aybah	1807 aytes	1857 azged	1907 aztij	1957 bafeg
1508 avime	1558 avvey	1608 awibo	1658 awxan	1708 axmev	1758 aybey	1808 aytic	1858 azgno	1908 aztoc	1958 bafib
1509 avinu	1559 avvim	1609 awink	1659 awxor	1709 axmim	1759 aybot	1809 aytog	1859 azgry	1909 aztun	1959 bafom
1510 avipy	**1560** avvok	**1610** awixu	**1660** awyac	**1710** axmon	**1760** aycaj	**1810** aytym	**1860** azguz	**1910** aztyd	**1960** bafru
1511 aviri	1561 avvuf	1611 awjip	1661 awyol	1711 axmuk	1761 aycos	1811 ayuch	1861 azhax	1911 azugi	1961 bagec
1512 avixo	1562 avwag	1612 awjok	1662 awzak	1712 axnan	1762 aydat	1812 ayudu	1862 azhub	1912 azunu	1962 bagif
1513 avjaw	1563 avweb	1613 awjun	1663 awzem	1713 axney	1763 aydru	1813 ayums	1863 azick	1913 azupo	1963 bagol
1514 avkef	1564 avwin	1614 awkah	1664 awziz	1714 axnum	1764 ayerk	1814 ayunk	1864 azidy	1914 azurb	1964 bagub
1515 avkoz	1565 avwow	1615 awket	1665 awzob	1715 axogi	1765 ayfal	1815 ayuto	1865 azifi	1915 azusa	1965 bahav
1516 avkru	1566 avwry	1616 awkow	1666 awzup	1716 axomy	1766 ayfet	1816 ayuve	1866 azips	1916 azute	1966 bahit
1517 avlah	1567 avxal	1617 awkri	1667 axade	1717 axono	1767 ayfif	1817 ayuxa	1867 azire	1917 azveh	1967 bahoy
1518 avlen	1568 avyaf	1618 awkyx	1668 axaku	1718 axork	1768 ayfoc	1818 ayvex	1868 azita	1918 azvip	1968 bahul
1519 avlos	1569 avyed	1619 awlaz	1669 axbaz	1719 axosa	1769 aygeh	1819 ayvig	1869 azker	1919 azvow	1969 baibe
1520 avluw	**1570** avyop	**1620** awlic	**1670** axbok	**1720** axovu	**1770** aygow	**1820** aywab	**1870** azkig	**1920** azvus	**1970** baica
1521 avmam	1571 avzad	1621 awlyl	1671 axbre	1721 axpes	1771 aygra	1821 aywen	1871 azkup	1921 azvyt	1971 baifo
1522 avmel	1572 avzig	1622 awmay	1672 axbus	1722 axpiz	1772 ayhag	1822 aywil	1872 azkyf	1922 azwav	1972 baigu
1523 avmir	1573 avzol	1623 awmis	1673 axbyt	1723 axpuc	1773 ayhem	1823 aywox	1873 azlem	1923 azwec	1973 baimp
1524 avmot	1574 avzub	1624 awnex	1674 axcel	1724 axpyd	1774 ayhiz	1824 ayxas	1874 azlin	1924 azwon	1974 bajaf
1525 avmud	1575 awaec	1625 awnil	1675 axcis	1725 axrar	1775 ayhyb	1825 ayxer	1875 azlob	1925 azxol	1975 bajed
1526 avnar	1576 awaha	1626 awnoy	1676 axcyp	1726 axrit	1776 ayilu	1826 azals	1876 azluf	1926 azzid	1976 bajoc
1527 avnez	1577 awain	1627 awnub	1677 axduy	1727 axryx	1777 ayins	1827 azamu	1877 azlyx	1927 azzle	1977 bakag
1528 avnog	1578 awalp	1628 awofa	1678 axeka	1728 axsas	1778 ayipa	1828 azank	1878 azmal	1928 baaba	1978 bakeb
1529 avnux	1579 awamb	1629 awoge	1679 axelb	1729 axsek	1779 ayixi	1829 azapi	1879 azmen	1929 baach	1979 bakil
1530 avnyk	**1580** awark	**1630** awolm	**1680** axenu	**1730** axsor	**1780** aykik	**1830** azart	**1880** azmos	**1930** baaju	**1980** bakun
1531 avoco	1581 awayo	1631 awopo	1681 axfec	1731 axspi	1781 aykof	1831 azavo	1881 azmyr	1931 baaly	1981 bakyd
1532 avohan	1582 awbaw	1632 aworn	1682 axflo	1732 axsyl	1782 aykux	1832 azawa	1882 aznam	1932 baant	1982 balef
1533 avoki	1583 awboj	1633 awosk	1683 axfub	1733 axtac	1783 aylak	1833 azaxe	1883 aznel	1933 baase	1983 balig
1534 avolt	1584 awcat	1634 awozu	1684 axgir	1734 axtil	1784 aylev	1834 azbaj	1884 aznit	1934 baazo	1984 baloz
1535 avons	1585 awcen	1635 awpef	1685 axgle	1735 axtof	1785 aylor	1835 azbil	1885 aznud	1935 babax	1985 baluv
1536 avowe	1586 awclo	1636 awpru	1686 axhah	1736 axugo	1786 aymaw	1836 azboz	1886 aznyc	1936 baber	1986 bamak
1537 avpan	1587 awcuf	1637 awraj	1687 axheg	1737 axuli	1787 aynad	1837 azbys	1887 azoko	1937 bablo	1987 bamem
1538 avpet	1588 awcyr	1638 awrey	1688 axhol	1738 axush	1788 aynir	1838 azcan	1888 azoma	1938 babup	1988 bamip
1539 avphy	1589 awdew	1639 awrix	1689 axhur	1739 axuta	1789 aynul	1839 azcey	1889 azope	1939 babyn	1989 bamob
1540 avpib	**1590** awdir	**1640** awser	**1690** axiby	**1740** axvav	**1790** ayoph	**1840** azcod	**1890** azosy	**1940** bacaz	**1990** bamuf
1541 avpod	1591 awdot	1641 awsid	1691 axico	1741 axvet	1791 ayout	1841 azcum	1891 azoti	1941 bacep	1991 bamyg
1542 avpum	1592 awdum	1642 awsky	1692 axifa	1742 axvif	1792 ayova	1842 azdaw	1892 azpas	1942 bacro	1992 banek
1543 avrac	1593 awdys	1643 awspa	1693 axils	1743 axvoy	1793 ayozi	1843 azdex	1893 azpek	1943 badev	1993 banix
1544 avreg	1594 aweal	1644 awsto	1694 axitu	1744 axwom	1794 aypew	1844 azdis	1894 azpix	1944 badic	1994 banyo
1545 avrom	1595 aweco	1645 awsuh	1695 axive	1745 axyca	1795 aypha	1845 azdly	1895 azpor	1945 badon	1995 baobs
1546 avrus	1596 aweig	1646 awtar	1696 axizi	1746 axzaf	1796 aypis	1846 azdur	1896 azrap	1946 badry	1996 baoda
1547 avsat	1597 awers	1647 awtel	1697 axjef	1747 axzly	1797 ayple	1847 azelf	1897 azrof	1947 baduk	1997 baofi
1548 avsoy	1598 aweux	1648 awtif	1698 axjid	1748 axzoc	1798 aypoy	1848 azene	1898 azruc	1948 baebu	1998 baojo
1549 avsul	1599 awfaf	1649 awtom	1699 axjot	1749 axzuz	1799 ayram	1849 azeva	1899 azryb	1949 baefa	1999 baoku

baowl bapam bapel bapiv

[redacted]

SECTION A

2000 bapot	2050 becyg	2100 beord	2150 bicad	2200 biogy	2250 bocux	2300 boost	2350 buepe	2400 bupoz	2450 byjor
2001 bapud	2051 beder	2101 beote	2151 biceo	2201 biols	2251 bodas	2301 booxy	2351 buert	2401 buput	2451 bykav
2002 baran	2052 bediz	2102 beoxo	2152 bicol	2202 biond	2252 bodek	2302 bopag	2352 buezu	2402 burak	2452 bykew
2003 barew	2053 bedov	2103 bepak	2153 bicri	2203 biorc	2253 bodor	2303 bopeb	2353 bufax	2403 burem	2453 bykiz
2004 barik	2054 beemi	2104 bepem	2154 bicub	2204 bioto	2254 bodug	2304 bopim	2354 bufer	2404 burig	2454 bykry
2005 barod	2055 beepa	2105 beplo	2155 bicyn	2205 bioze	2255 bodyc	2305 bopow	2355 bufid	2405 burob	2455 byler
2006 barum	2056 beero	2106 bepry	2156 bidaf	2206 bipar	2256 boelz	2306 bopun	2356 bugay	2406 buruf	2456 bylup
2007 baryl	2057 beest	2107 bepuf	2157 bidil	2207 bipez	2257 boemy	2307 boraw	2357 bugep	2407 busal	2457 bymaz
2008 basap	2058 befaz	2108 beral	2158 bidyx	2208 bipik	2258 boept	2308 boref	2358 bugil	2408 buscu	2458 bym[illegible]in
2009 bases	2059 befep	2109 beren	2159 bieba	2209 bipog	2259 boere	2309 boroz	2359 buhaz	2409 buski	2459 bymok
2010 basiz	2060 befin	2110 berij	2160 biech	2210 bipyt	2260 boeso	2310 bosah	2360 buhex	2410 busos	2460 by[illegible]ev
2011 basof	2061 befly	2111 beros	2161 biedo	2211 biras	2261 boevu	2311 bosem	2361 buhik	2411 butap	2461 bynig
2012 basuc	2062 befok	2112 berut	2162 biege	2212 birek	2262 bofav	2312 bosuf	2362 buhum	2412 butew	2462 bynuk
2013 basyr	2063 befus	2113 besam	2163 biemp	2213 birib	2263 bofey	2313 botep	2363 buhyp	2413 butiz	2463 bynyp
2014 bataw	2064 begev	2114 besel	2164 bient	2214 biror	2264 bofiz	2314 botiv	2364 buide	2414 butur	2464 byody
2015 batex	2065 begis	2115 b[illegible]sho	2165 biesu	2215 birug	2265 bofop	2315 botot	2365 buili	2415 buvam	2465 byola
2016 bator	2066 begon	2116 besif	2166 bifag	2216 bisav	2266 bofur	2316 botul	2366 buimu	2416 buvel	2466 byomo
2017 batuy	2067 begyr	2117 besud	2167 bifeb	2217 bisly	2267 boger	2317 boudy	2367 buina	2417 buvij	2467 byong
2018 bauga	2068 behap	2118 besyk	2168 bifir	2218 bisop	2268 bogib	2318 boval	2368 buipo	2418 buvot	2468 byose
2019 baumi	2069 behew	2119 betat	2169 bifox	2219 bisur	2269 bogov	2319 bovit	2369 buist	2419 buwan	2469 bypec
2020 baurn	2070 behof	2120 betez	2170 bifun	2220 bithy	2270 bohaj	2320 bovos	2370 bujab	2420 buzeh	2470 bypij
2021 bautu	2071 behuc	2121 betig	2171 bigat	2221 bitit	2271 bohir	2321 boway	2371 bujif	2421 buzof	2471 bypol
2022 bavar	2072 behyd	2122 betoy	2172 bigef	2222 bitow	2272 bohox	2322 bowex	2372 bujon	2422 byaci	2472 bypub
2023 bavog	2073 beibs	2123 betru	2173 bigid	2223 bitre	2273 bohuk	2323 boxin	2373 bujuk	2423 byafu	2473 bypyr
2024 bawat	2074 beipi	2124 beufo	2174 bigoz	2224 biube	2274 bohys	2324 boyaz	2374 bukac	2424 bybel	2474 byraf
2025 bawid	2075 beirl	2125 beuma	2175 bigul	2225 biuja	2275 boics	2325 bozan	2375 bukeg	2425 bybiv	2475 byril
2026 bayah	2076 bejuw	2126 beund	2176 bigyb	2226 biult	2276 boigo	2326 bozum	2376 bukit	2426 bycac	2476 byruz
2027 bayen	2077 bekad	2127 beuti	2177 bihaw	2227 biumy	2277 boika	2327 buanz	2377 bukul	2427 byceg	2477 bysox
2028 bayow	2078 bekec	2128 bevan	2178 bihut	2228 biunz	2278 boiph	2328 buapa	2378 bukyn	2428 bycys	2478 bysyd
2029 bayut	2079 bekir	2129 bevik	2179 bihyc	2229 biver	2279 boity	2329 buask	2379 bulad	2429 bydan	2479 bytah
2030 bazet	2080 bekol	2130 bevum	2180 bijak	2230 bivov	2280 boizu	2330 buath	2380 bulec	2430 bydey	2480 bytim
2031 bazop	2081 bekub	2131 bewar	2181 bijem	2231 bivup	2281 bojat	2331 buaxo	2381 buliv	2431 bydif	2481 bytob
2032 bazur	2082 beled	2132 bewes	2182 bijob	2232 biwaz	2282 bojew	2332 buaye	2382 bulol	2432 bydod	2482 bytyn
2033 beabu	2083 belip	2133 bewot	2183 bijuf	2233 biwet	2283 bokab	2333 bubar	2383 bulub	2433 bydut	2483 byuco
2034 beaca	2084 beloc	2134 bezey	2184 bikal	2234 biwod	2284 bokev	2334 bubez	2384 bumaf	2434 byeja	2484 byuda
2035 beafi	2085 beluz	2135 bezul	2185 biken	2235 bizah	2285 bokis	2335 bubly	2385 bumed	2435 byeli	2485 byuky
2036 beago	2086 belyn	2136 biabo	2186 bikyk	2236 bizin	2286 boklu	2336 bubog	2386 bumoc	2436 byerx	2486 byume
2037 bealm	2087 bemag	2137 biacs	2187 bilam	2237 boahs	2287 bokon	2337 bubux	2387 bumuz	2437 byews	2487 byuph
2038 beane	2088 bemeb	2138 biald	2188 bilel	2238 boaki	2288 bokyr	2338 bucas	2388 bunag	2438 byfro	2488 byuvi
2039 beaty	2089 bemil	2139 biapy	2189 bilon	2239 boalb	2289 bolac	2339 bucin	2389 buneb	2439 byful	2489 byuzu
2040 beawk	2090 bemox	2140 biarm	2190 bilud	2240 boamt	2290 boleg	2340 bucor	2390 bunis	2440 bygez	2490 byvay
2041 bebas	2091 bemun	2141 biate	2191 biman	2241 boawe	2291 bolom	2341 bucug	2391 bunun	2441 byhib	2491 byvef
2042 bebek	2092 bemyc	2142 biawn	2192 bimex	2242 boaya	2292 bolyp	2342 bucyb	2392 buobi	2442 byhov	2492 byvoz
2043 bebli	2093 benah	2143 biaxi	2193 bimig	2243 bobap	2293 bomix	2343 budav	2393 buodu	2443 byhyg	2493 bywaw
2044 bebor	2094 benef	2144 biaza	2194 bimoh	2244 bobes	2294 bonaf	2344 budej	2394 buoft	2444 byiba	2494 caabe
2045 bebug	2095 benib	2145 bibac	2195 bimum	2245 bobof	2295 bonoc	2345 budib	2395 buogo	2445 byife	2495 caack
2046 becav	2096 benoz	2146 bibeg	2196 bimyl	2246 bobuc	2296 bonuz	2346 budop	2396 buoha	2446 byimi	2496 caaft
2047 becit	2097 benyl	2147 bibif	2197 binap	2247 bochy	2297 booca	2347 buebs	2397 buoys	2447 byiru	2497 caagi
2048 becop	2098 beoci	2148 biblu	2198 binic	2248 bocic	2298 boodi	2348 buegi	2398 bupaw	2448 byiso	2498 caahu
2049 becur	2099 beoky	2149 bibro	2199 biobu	2249 bocog	2299 boonu	2349 buema	2399 bupir	2449 byjas	2499 caaja

2500 caamy	**2550** cakok	**2600** cazel	**2650** cejuk	**2700** cezof	**2750** cimud	**2800** coaxi	**2850** cokal	**2900** cuaho	**2950** cukes
2501 caarn	2551 cakus	2601 cazoy	2651 cekac	2701 ciacu	2751 cinar	2801 coaza	2851 coken	2901 cualo	2951 cukim
2502 caasp	2552 cakym	2602 ceami	2652 cekeg	2702 ciams	2752 cinez	2802 cobac	2852 cokos	2902 cuana	2952 cukuc
2503 caawo	2553 calal	2603 ceanz	2653 cekit	2703 ciaro	2753 cinip	2803 cobeg	2853 cokyk	2903 cuari	2953 cular
2504 cabet	2554 calib	2604 ceapa	2654 cekul	2704 ciata	2754 cinog	2804 cobif	2854 colam	2904 cuasm	2954 culez
2505 cabiz	2555 caloh	2605 ceask	2655 cekyn	2705 ciave	2755 cinux	2805 coblu	2855 colel	2905 cuaxu	2955 culid
2506 cabod	2556 calut	2606 ceath	2656 celad	2706 ciaxy	2756 cinyk	2806 cocad	2856 colix	2906 cubag	2956 culog
2507 cabum	2557 calyr	2607 ceaxo	2657 celec	2707 ciazi	2757 cioco	2807 cocec	2857 colud	2907 cubef	2957 culux
2508 cabyc	2558 camac	2608 ceaye	2658 celiv	2708 cibab	2758 ciohm	2808 cocol	2858 comex	2908 cubib	2958 culym
2509 cacap	2559 cameg	2609 cebar	2659 celol	2709 cibev	2759 ciolt	2809 cocri	2859 comoh	2909 cubun	2959 cumas
2510 caces	**2560** camif	**2610** cebez	**2660** celub	**2710** cibis	**2760** cions	**2810** cocub	**2860** comyl	**2910** cubyl	**2960** cumek
2511 caciv	2561 camom	2611 cebim	2661 cemaf	2711 cibon	2761 ciowe	2811 cocyn	2861 conap	2911 cucay	2961 cumit
2512 cacof	2562 canew	2612 cebly	2662 cemed	2712 cibuk	2762 cipan	2812 codaf	2862 conot	2912 cucet	2962 cumor
2513 cacuc	2563 caobo	2613 cebog	2663 cemoc	2713 cicem	2763 cipet	2813 codil	2863 coobu	2913 cucoz	2963 cumug
2514 caeyx	2564 caofy	2614 cebux	2664 cemuz	2714 ciciz	2764 ciphy	2814 codoc	2864 coogy	2914 cudak	2964 cunav
2515 cadar	2565 caome	2615 cecas	2665 cenag	2715 cicra	2765 cipib	2815 coduz	2865 coond	2915 cudem	2965 cuneh
2516 cadez	2566 caonc	2616 cecin	2666 ceneb	2716 cicyc	2766 cipod	2816 codyx	2866 coorc	2916 cudig	2966 cunop
2517 cadip	2567 caopa	2617 cecor	2667 cenis	2717 cidap	2767 cipum	2817 coeba	2867 cooto	2917 cudob	2967 cunur
2518 cadog	2568 caori	2618 cecug	2668 cenox	2718 cides	2768 cirac	2818 coech	2868 cooze	2918 cuduf	2968 cuobe
2519 cadux	2569 caots	2619 cecyb	2669 cenun	2719 cidof	2769 cireg	2819 coedo	2869 copar	2919 cudyn	2969 cuock
2520 caedy	**2570** capaf	**2620** cedav	**2670** ceobi	**2720** ciduc	**2770** cirus	**2820** coege	**2870** copez	**2920** cuebo	**2970** cuolo
2521 caefo	2571 caped	2621 cedej	2671 ceodu	2721 cielu	2771 cisoy	2821 coeki	2871 copik	2921 cuede	2971 cuomu
2522 caelv	2572 capil	2622 cedib	2672 ceoft	2722 cieny	2772 cisul	2822 coemp	2872 copog	2922 cuefy	2972 cuoni
2523 caemu	2573 carag	2623 cedop	2673 ceogo	2723 ciepo	2773 citid	2823 coent	2873 copyt	2923 cueku	2973 cuopy
2524 caerg	2574 careb	2624 ceebs	2674 ceoha	2724 ciets	2774 citys	2824 coesu	2874 corek	2924 cuela	2974 cuota
2525 caese	2575 carid	2625 ceegi	2675 ceove	2725 ciexi	2775 ciufu	2825 cofag	2875 corib	2925 cuern	2975 cupev
2526 cafas	2576 carun	2626 ceema	2676 ceoys	2726 cieze	2776 ciujo	2826 cofeb	2876 corug	2926 cueti	2976 cupif
2527 cafix	2577 caryf	2627 ceeno	2677 cerak	2727 cifay	2777 ciulk	2827 cofir	2877 cosav	2927 cufaw	2977 cupuk
2528 cafor	2578 casaw	2628 ceepe	2678 cerem	2728 cifex	2778 civaz	2828 cofox	2878 cosly	2928 cufen	2978 cupyc
2529 cafug	2579 casef	2629 ceert	2679 cerob	2729 cifil	2779 civey	2829 cofun	2879 cosur	2929 cufoy	2979 curov
2530 cagav	**2580** casit	**2630** ceezu	**2680** ceruf	**2730** cifoh	**2780** civim	**2830** cogat	**2880** cothy	**2930** cufut	**2980** cusaz
2531 cagim	2581 casoz	2631 cefax	2681 cesal	2731 cifuz	2781 civok	2831 cogef	2881 cotit	2931 cugam	2981 cusep
2532 cagop	2582 casuv	2632 cefer	2682 cesen	2732 cigas	2782 civuf	2832 cogid	2882 cotow	2932 cugel	2982 cusmi
2533 cagur	2583 casys	2633 cefid	2683 ceski	2733 cigek	2783 ciwag	2833 cogoz	2883 cotre	2933 cugip	2983 cusow
2534 cagyn	2584 caten	2634 cegay	2684 cesos	2734 cigit	2784 ciweb	2834 cogul	2884 coube	2934 cugot	2984 custy
2535 cahab	2585 catry	2635 cegep	2685 cetap	2735 cigor	2785 ciwin	2835 cogyb	2885 couja	2935 cugru	2985 cusus
2536 cahek	2586 catuk	2636 cegil	2686 cetew	2736 cigug	2786 ciwow	2836 cohaw	2886 coult	2936 cuhaf	2986 cutal
2537 cahig	2587 cauba	2637 cegus	2687 cetiz	2737 cihec	2787 ciwry	2837 cohok	2887 coumy	2937 cuher	2987 cuteg
2538 cahoa	2588 caufe	2638 cehaz	2688 cetod	2738 cihiv	2788 cixal	2838 cohut	2888 counz	2938 cuhis	2988 cutin
2539 cahup	2589 cauku	2639 cehex	2689 cetur	2739 cikef	2789 cixur	2839 cohyc	2889 covov	2939 cuhod	2989 cutum
2540 caimo	**2590** cauni	**2640** cehik	**2690** ceuca	**2740** cikoz	**2790** cizad	**2840** coibi	**2890** covup	**2940** cuhuz	**2990** cutyx
2541 caihy	2591 cauzo	2641 cehum	2691 ceudo	2741 cikru	2791 cizol	2841 coicu	2891 cowaz	2941 cuidi	2991 cuvad
2542 caipu	2592 cavak	2642 cehyp	2692 ceunt	2742 cilah	2792 cizub	2842 coida	2892 cowet	2942 cuifu	2992 cuvec
2543 caiti	2593 cavic	2643 ceide	2693 ceure	2743 cilen	2793 coabo	2843 coiko	2893 cowod	2943 cuilk	2993 cuvir
2544 caiwa	2594 cavul	2644 ceimu	2694 cevel	2744 cilos	2794 coacs	2844 coing	2894 coyew	2944 cuite	2994 cuvol
2545 cajah	2595 caweh	2645 ceina	2695 cevij	2745 ciluw	2795 coald	2845 coisy	2895 coyla	2945 cuivy	2995 cuwac
2546 cajis	2596 cayle	2646 ceist	2696 cevot	2746 cimam	2796 coapy	2846 cojak	2896 cozah	2946 cujan	2996 cuwed
2547 cakaz	2597 cayot	2647 cejab	2697 cewan	2747 cimel	2797 coarm	2847 cojem	2897 cozin	2947 cujeb	2997 cuyls
2548 cakep	2598 cayub	2648 cejif	2698 ceyat	2748 cimir	2798 coate	2848 cojob	2898 cuaby	2948 cujik	2998 cuyoc
2549 cakin	2599 cazam	2649 cejon	2699 cezeh	2749 cimot	2799 coawn	2849 cojuf	2899 cuadz	2949 cukap	2999 cuzat

cyahs cyaki cyalb cyamt

SECTION A

3000 cyaud	**3050** cyleg	**3100** dabac	**3150** dakal	**3200** dawet	**3250** dekis	**3300** dialo	**3350** disli	**3400** doele	**3450** doowi
3001 cyawe	3051 cylif	3101 dabeg	3151 daken	3201 dawod	3251 deklu	3301 diamp	3351 disov	3401 doens	3451 dopam
3002 cyaya	3052 cylom	3102 dabif	3152 dakos	3202 dayew	3252 dekyr	3302 dianu	3352 dispu	3402 doeph	3452 dopel
3003 cybap	3053 cylyp	3103 dablu	3153 dakyk	3203 dayla	3253 dekon	3303 diard	3353 disyc	3403 doerd	3453 dopiv
3004 cybes	3054 cymad	3104 dabro	3154 dalam	3204 dazah	3254 delac	3304 diasy	3354 ditaf	3404 doewi	3454 dopot
3005 cybig	3055 cymec	3105 dacad	3155 dalel	3205 dazin	3255 deleg	3305 diaxt	3355 dites	3405 dofac	3455 dopud
3006 cybof	3056 cymix	3106 dacec	3156 dalix	3206 deahs	3256 delif	3306 dibah	3356 ditic	3406 dofeg	3456 dorew
3007 cybuc	3057 cynaf	3107 dacol	3157 dalon	3207 deaki	3257 delom	3307 dibey	3357 ditym	3407 dofib	3457 dosap
3008 cybyb	3058 cyned	3108 dacri	3158 dalud	3208 dealb	3258 delyp	3308 dibot	3358 diuch	3408 dofom	3458 dosiz
3009 cycic	3059 cynoc	3109 dacub	3159 damex	3209 deamt	3259 demad	3309 dicaj	3359 diudu	3409 dofru	3459 dosof
3010 cydas	**3060** cynuz	**3110** dacyn	**3160** damig	**3210** deawe	**3260** demec	**3310** dicos	**3360** diums	**3410** dogec	**3460** dosuc
3011 cydek	3061 cyoca	3111 dadaf	3161 damoh	3211 deaya	3261 demol	3311 didat	3361 diunk	3411 dogif	3461 dosyr
3012 cydor	3062 cyodi	3112 daded	3162 damum	3212 debap	3262 denaf	3312 didru	3362 diuri	3412 dogol	3462 dotaw
3013 cydug	3063 cyoho	3113 dadil	3163 damyl	3213 debes	3263 dened	3313 dierk	3363 diuto	3413 dogub	3463 dotex
3014 cydyc	3064 cyonu	3114 dadoc	3164 danap	3214 debig	3264 denoc	3314 dievo	3364 diuxa	3414 dohav	3464 dotuy
3015 cyeid	3065 cyost	3115 daduz	3165 danot	3215 debof	3265 denuz	3315 difal	3365 divex	3415 dohit	3465 douga
3016 cyelz	3066 cyoxy	3116 dadyx	3166 danic	3216 debuc	3266 deoca	3316 difet	3366 divig	3416 dohoy	3466 doumi
3017 cyemy	3067 cypag	3117 daeba	3167 daobu	3217 debyb	3267 deodi	3317 difif	3367 divyn	3417 dohul	3467 dourn
3018 cyept	3068 cypeb	3118 daech	3168 daogy	3218 decar	3268 deoho	3318 difoc	3368 diwab	3418 doibe	3468 doutu
3019 cyere	3069 cypim	3119 daedo	3169 daols	3219 dechy	3269 deonu	3319 digeh	3369 diwen	3419 doica	3469 dovar
3020 cyeso	**3070** cypow	**3120** daege	**3170** daond	**3220** decic	**3270** deost	**3320** digow	**3370** diwil	**3420** doifo	**3470** dovez
3021 cyevu	3071 cypun	3121 daeki	3171 daorc	3221 decog	3271 deoxy	3321 digra	3371 diwox	3421 doigu	3471 dovin
3022 cyfav	3072 cyraw	3122 daemp	3172 daoto	3222 decux	3272 depag	3322 dihem	3372 dixas	3422 doimp	3472 dovog
3023 cyfey	3073 cyref	3123 daent	3173 daoze	3223 dedas	3273 depeb	3323 dihiz	3373 dixer	3423 dojaf	3473 dowat
3024 cyfiz	3074 cyroz	3124 daesu	3174 dapar	3224 dedek	3274 depim	3324 dihyb	3374 dizac	3424 dojed	3474 dowid
3025 cyfop	3075 cysah	3125 dafag	3175 dapez	3225 dedug	3275 depow	3325 dijaz	3375 dizep	3425 dojoc	3475 doyah
3026 cyfur	3076 cysem	3126 dafeb	3176 dapik	3226 dedyc	3276 dep[illegible]	3326 dijec	3376 dizon	3426 dokag	3476 dozet
3027 cygax	3077 cysip	3127 dafir	3177 dapog	3227 deelz	3277 deraw	3327 dijud	3377 dizuf	3427 dokeb	3477 dozop
3028 cyger	3078 cysob	3128 dafox	3178 dapyt	3228 deept	3278 deref	3328 dikof	3378 doaba	3428 dokil	3478 dozur
3029 cygib	3079 cysuf	3129 dafun	3179 daras	3229 deere	3279 deroz	3329 dikux	3379 doach	3429 dokun	3479 duacu
3030 cygov	**3080** cytam	**3130** dagat	**3180** darek	**3230** deevu	**3280** desah	**3330** dilak	**3380** doaim	**3430** dokyd	**3480** duams
3031 cygup	3081 cytep	3131 dagef	3181 darib	3231 defav	3281 desem	3331 dilev	3381 doaju	3431 dolef	3481 duaro
3032 cyhaj	3082 cytiv	3132 dagid	3182 daror	3232 defey	3282 desip	3332 dilor	3382 doaly	3432 dolig	3482 duata
3033 cyhir	3083 cytul	3133 dagoz	3183 darug	3233 defiz	3283 desuf	3333 dimaw	3383 doant	3433 doloz	3483 duave
3034 cyhox	3084 cyudy	3134 dagul	3184 dasav	3234 defop	3284 detam	3334 dimyx	3384 doase	3434 doluv	3484 duaxy
3035 cyhuk	3085 cyval	3135 dagyb	3185 dasim	3235 defur	3285 detep	3335 dinir	3385 doazo	3435 domak	3485 duazi
3036 cyhys	3086 cyven	3136 dahaw	3186 dasly	3236 degib	3286 detiv	3336 dinul	3386 dobax	3436 domem	3486 dubev
3037 cyics	3087 cyvit	3137 dahok	3187 dasop	3237 degov	3287 detot	3337 diony	3387 dober	3437 domip	3487 dubon
3038 cyigo	3088 cyvos	3138 dahut	3188 dasur	3238 dehaj	3288 detul	3338 dioph	3388 dobup	3438 domob	3488 dubuk
3039 cyika	3089 cyway	3139 dahyc	3189 dathy	3239 dehir	3289 deudy	3339 diore	3389 dobyn	3439 domuf	3489 duciz
3040 cyiph	**3090** cywex	**3140** daibi	**3190** datow	**3240** dehox	**3290** deval	**3340** diova	**3390** docaz	**3440** domyg	**3490** ducob
3041 cyity	3091 daabo	3141 daicu	3191 datre	3241 dehuk	3291 deven	3341 diozi	3391 docep	3441 donal	3491 ducra
3042 cyizu	3092 daacs	3142 daida	3192 dauja	3242 deics	3292 devit	3342 dipew	3392 docro	3442 donek	3492 ducyc
3043 cykab	3093 daald	3143 daiko	3193 dault	3243 deigo	3293 devos	3343 dipha	3393 dodab	3443 donix	3493 dudap
3044 cykev	3094 daapy	3144 daing	3194 daumy	3244 deika	3294 deway	3344 dipoy	3394 dodev	3444 donyo	3494 dudof
3045 cykis	3095 daarm	3145 daith	3195 daunz	3245 deizu	3295 dewex	3345 diram	3395 dodic	3445 doobs	3495 duduc
3046 cyklu	3096 daate	3146 dajak	3196 daver	3246 dejat	3296 dezan	3346 direl	3396 dodon	3446 dooda	3496 dueha
3047 cykon	3097 daawn	3147 dajem	3197 davov	3247 dejew	3297 dezik	3347 diroj	3397 dodry	3447 doofi	3497 duelu
3048 cykyr	3098 daaxi	3148 dajob	3198 davup	3248 dekab	3298 dezum	3348 diruw	3398 doduk	3448 doojo	3498 dueny
3049 cylac	3099 daaza	3149 dajuf	3199 dawaz	3249 dekev	3299 diaka	3349 diske	3399 doebu	3449 dooku	3499 duepo

3500 duexi	3550 dupum	3600 dydej	3650 dyogo	3700 ebcyb	3750 ebnox	3800 eccax	3850 ecnay	3900 edaty	3950 edmil
3501 dueze	3551 durac	3601 dydib	3651 dyoha	3701 ebdav	3751 ebnun	3801 eccer	3851 ecnem	3901 edawk	3951 edmox
3502 dufay	3552 dureg	3602 dydop	3652 dyove	3702 ebdej	3752 ebobi	3802 eccib	3852 ecnid	3902 edbas	3952 edmun
3503 dufex	3553 durom	3603 dyebs	3653 dyoys	3703 ebdib	3753 ebodu	3803 eccov	3853 ecnob	3903 edbek	3953 edmyc
3504 dufil	3554 dusat	3604 dyegi	3654 dypaw	3704 ebdop	3754 eboft	3804 eccup	3854 ecnuf	3904 edbli	3954 ednah
3505 dufoh	3555 dusoy	3605 dyema	3655 dypir	3705 ebebs	3755 ebogo	3805 ecdaz	3855 ecoak	3905 edbor	3955 ednef
3506 dufuz	3556 dusul	3606 dyeno	3656 dypoz	3706 ebegi	3756 eboha	3806 ecdep	3856 ecofu	3906 edbug	3956 ednib
3507 dugas	3557 dutax	3607 dyepe	3657 dyput	3707 ebema	3757 ebove	3807 ecdix	3857 ecoly	3907 edcav	3957 ednoz
3508 dugek	3558 duter	3608 dyert	3658 dyrak	3708 ebeno	3758 eboys	3808 ecdok	3858 ecomp	3908 edcit	3958 ednyl
3509 dugit	3559 dutid	3609 dyezu	3659 dyrem	3709 ebepe	3759 ebpaw	3809 ecdus	3859 econe	3909 edcop	3959 edoba
3510 dugor	3560 dutup	3610 dyfax	3660 dyrig	3710 ebert	3760 ebpir	3810 ecedi	3860 ecort	3910 edcur	3960 edoci
3511 dugug	3561 dutys	3611 dyfer	3661 dyrob	3711 ebezu	3761 ebpoz	3811 ecemo	3861 ecosi	3911 edcyg	3961 edoky
3512 duhec	3562 duvaz	3612 dyfid	3662 dyruf	3712 ebfax	3762 ebput	3812 eceng	3862 ecowo	3912 eddiz	3962 edord
3513 duhiv	3563 duvey	3613 dygay	3663 dysal	3713 ebfer	3763 ebrak	3813 ecerl	3863 ecpal	3913 eddov	3963 edote
3514 duids	3564 duvim	3614 dygep	3664 dyscu	3714 ebfid	3764 ebrem	3814 ecesy	3864 ecpen	3914 edepa	3964 edown
3515 duiga	3565 duvok	3615 dygil	3665 dysen	3715 ebgay	3765 ebrig	3815 ecewa	3865 ecpig	3915 edero	3965 edoxo
3516 duime	3566 duvuf	3616 dygus	3666 dyski	3716 ebgep	3766 ebrob	3816 ecexe	3866 ecpos	3916 edewe	3966 edpak
3517 duinu	3567 duwag	3617 dyhaz	3667 dysos	3717 ebgil	3767 ebruf	3817 ecfab	3867 ecrat	3917 edfaz	3967 edpam
3518 duipy	3568 duweb	3618 dyhex	3668 dytap	3718 ebgus	3768 ebsal	3818 ecfev	3868 ecrhi	3918 edfep	3968 edpid
3519 duiri	3569 duwin	3619 dyhik	3669 dytew	3719 ebhaz	3769 ebscu	3819 ecfip	3869 ecryn	3919 edfin	3969 edplo
3520 duixo	3570 duwow	3620 dyhum	3670 dytiz	3720 ebhex	3770 ebsen	3820 ecflu	3870 ecsew	3920 edfly	3970 edpry
3521 dujaw	3571 duwry	3621 dyhyp	3671 dytod	3721 ebhik	3771 ebski	3821 ecfon	3871 ecsil	3921 edfok	3971 edpuf
3522 dukef	3572 duxal	3622 dyide	3672 dytur	3722 ebhum	3772 ebsos	3822 ecgac	3872 ecsod	3922 edfus	3972 edral
3523 dukoz	3573 duxur	3623 dyili	3673 dytyl	3723 ebhyp	3773 ebtap	3823 ecgeg	3873 ecsum	3923 edgab	3973 edren
3524 dukru	3574 duyaf	3624 dyimu	3674 dyuca	3724 ebide	3774 ebtew	3824 ecgin	3874 ectas	3924 edgev	3974 edrij
3525 dulah	3575 duyed	3625 dyina	3675 dyudo	3725 ebili	3775 ebtiz	3825 ecgom	3875 ectey	3925 edgon	3975 edros
3526 dulen	3576 duyop	3626 dyipo	3676 dyunt	3726 ebimu	3776 ebtod	3826 ecgyl	3876 ectik	3926 edhap	3976 edrut
3527 dulos	3577 duzad	3627 dyist	3677 dyure	3727 ebina	3777 ebtur	3827 echif	3877 ectol	3927 edhew	3977 edsam
3528 duluw	3578 duzig	3628 dykac	3678 dyvam	3728 ebipo	3778 ebtyl	3828 echot	3878 ectut	3928 edhof	3978 edsel
3529 dumam	3579 duzol	3629 dykeg	3679 dyvel	3729 ebist	3779 ebuca	3829 echug	3879 ecunc	3929 edhuc	3979 edsho
3530 dumel	3580 duzub	3630 dykit	3680 dyvij	3730 ebjab	3780 ebudo	3830 echyr	3880 ecurg	3930 edhyd	3980 edsif
3531 dumir	3581 dyami	3631 dykul	3681 dyvot	3731 ebjif	3781 ebunt	3831 ecigi	3881 ecust	3931 edice	3981 edsud
3532 dumot	3582 dyanz	3632 dykyn	3682 dywan	3732 ebjon	3782 ebure	3832 ecism	3882 ecuxo	3932 ediod	3982 edsyk
3533 dumud	3583 dyapa	3633 dylad	3683 ebami	3733 ebjuk	3783 ebvam	3833 ecjeh	3883 ecuze	3933 edipi	3983 edtat
3534 dunez	3584 dyask	3634 dylec	3684 ebanz	3734 ebkac	3784 ebvel	3834 ecjim	3884 ecvap	3934 edirl	3984 edtez
3535 dunip	3585 dyath	3635 dyliv	3685 ebapa	3735 ebkeg	3785 ebvij	3835 ecjoy	3885 ecves	3935 edjac	3985 edtig
3536 dunog	3586 dyauc	3636 dylol	3686 ebask	3736 ebkit	3786 ebvot	3836 ecjub	3886 ecviv	3936 edjeg	3986 edtoy
3537 dunux	3587 dyaxo	3637 dylub	3687 ebath	3737 ebkul	3787 ebwan	3837 eckaf	3887 ecxel	3937 edjuw	3987 edtru
3538 dunyk	3588 dybar	3638 dymaf	3688 ebauc	3738 ebkyn	3788 ebyat	3838 ecked	3888 ecyar	3938 edkad	3988 edufo
3539 duoco	3589 dybez	3639 dymed	3689 ebaxo	3739 eblad	3789 ebzeh	3839 eckli	3889 ecyex	3939 edkec	3989 eduma
3540 duohm	3590 dybim	3640 dymoc	3690 ebaye	3740 eblec	3790 ebzof	3840 eckoc	3890 ecyog	3940 edkir	3990 edund
3541 duoki	3591 dybly	3641 dymuz	3691 ebbar	3741 ebliv	3791 ecact	3841 eclag	3891 eczek	3941 edkol	3991 eduti
3542 duolt	3592 dybog	3642 dynag	3692 ebbez	3742 eblol	3792 ecady	3842 ecleb	3892 eczir	3942 edkub	3992 edvan
3543 duons	3593 dybux	3643 dyneb	3693 ebbly	3743 eblub	3793 ecage	3843 ecliz	3893 edabu	3943 edlaf	3993 edvik
3544 duowe	3594 dycas	3644 dynis	3694 ebbog	3744 ebmaf	3794 ecama	3844 eclox	3894 edaca	3944 edled	3994 edvum
3545 dupan	3595 dycin	3645 dynox	3695 ebbux	3745 ebmed	3795 ecapo	3845 eclun	3895 edads	3945 edloc	3995 edwar
3546 dupet	3596 dycor	3646 dynun	3696 ebcas	3746 ebmoc	3796 ecbav	3846 eclyt	3896 edafi	3946 edluz	3996 edwes
3547 duphy	3597 dycug	3647 dyobi	3697 ebcin	3747 ebmuz	3797 ecbij	3847 ecmah	3897 edago	3947 edlyn	3997 edwot
3548 dupib	3598 dycyb	3648 dyodu	3698 ebcor	3748 ebnag	3798 ecble	3848 ecmef	3898 edalm	3948 edmag	3998 edyaw
3549 dupod	3599 dydav	3649 dyoft	3699 ebcug	3749 ebneb	3799 ecboh	3849 ecmoz	3899 edane	3949 edmeb	3999 edzey

edzog edzul efahs efaki

[redacted]

SECTION A

4000 efalb	**4050** efkab	**4100** efxin	**4150** egigu	**4200** egtor	**4250** ehgod	**4300** ehted	**4350** ejlev	**4400** ekang	**4450** ekkod
4001 efamt	4051 efkev	4101 efzan	4151 egimp	4201 egtuy	4251 ehgum	4301 ehtip	4351 ejlor	4401 ekape	4451 ekkum
4002 efaud	4052 efkis	4102 efzik	4152 egits	4202 eguga	4252 ehici	4302 ehtox	4352 ejlys	4402 ekara	4452 eklap
4003 efawe	4053 efklu	4103 efzum	4153 egjaf	4203 egumi	4253 ehigy	4303 ehtuc	4353 ejmaw	4403 ekaul	4453 ekles
4004 efaya	4054 efkon	4104 egaba	4154 egjed	4204 egurn	4254 ehiku	4304 ehtyt	4354 ejmyx	4404 ekbaf	4454 eklof
4005 efbap	4055 efkyr	4105 egach	4155 egjoc	4205 egutu	4255 ehima	4305 ehubo	4355 ejnad	4405 ekbed	4455 ekluc
4006 efbes	4056 eflac	4106 egaim	4156 egkag	4206 egvar	4256 ehise	4306 ehuka	4356 ejnul	4406 ekbit	4456 ekmar
4007 efbig	4057 efleg	4107 egaju	4157 egkeb	4207 egvez	4257 ehjap	4307 ehulu	4357 ejony	4407 ekboc	4457 ekmez
4008 efbof	4058 eflif	4108 egaly	4158 egkil	4208 egvin	4258 ehjes	4308 ehune	4358 ejoph	4408 ekbry	4458 ekmid
4009 efbuc	4059 eflom	4109 egant	4159 egkun	4209 egvog	4259 ehjil	4309 ehvat	4359 ejore	4409 ekbuz	4459 ekmog
4010 efbyb	**4060** eflyp	**4110** egase	**4160** egkyd	**4210** egwat	**4260** ehjof	**4310** ehvix	**4360** ejoso	**4410** ekcag	**4460** ekmux
4011 efcar	4061 efmad	4111 egazo	4161 eglay	4211 egwid	4261 ehjux	4311 ehvoc	4361 ejout	4411 ekceb	4461 ekmys
4012 efchy	4062 efmec	4112 egbax	4162 eglef	4212 egyah	4262 ehkar	4312 ehwiv	4362 ejova	4412 ekcil	4462 eknet
4013 efcic	4063 efmix	4113 egber	4163 egloz	4213 egyen	4263 ehkib	4313 ehyeb	4363 ejozi	4413 ekcox	4463 eknor
4014 efcog	4064 efmol	4114 egblo	4164 egluv	4214 egyow	4264 ehkog	4314 ehzag	4364 ejpew	4414 ekcun	4464 eknug
4015 efcux	4065 efnaf	4115 egbup	4165 egmak	4215 egyut	4265 ehlas	4315 ejahi	4365 ejpha	4415 ekcyd	4465 eknyx
4016 efdas	4066 efned	4116 egbyn	4166 egmem	4216 egzet	4266 ehlek	4316 ejaka	4366 ejpis	4416 ekday	4466 ekoga
4017 efdek	4067 efnoc	4117 egcaz	4167 egmip	4217 egzop	4267 ehlim	4317 ejamp	4367 ejple	4417 ekdef	4467 ekolu
4018 efdor	4068 efnuz	4118 egcep	4168 egmob	4218 egzur	4268 ehlow	4318 ejanu	4368 ejpoy	4418 ekdim	4468 ekopi
4019 efdug	4069 efoca	4119 egcro	4169 egmuf	4219 ehacy	4269 ehlug	4319 ejard	4369 ejram	4419 ekdoz	4469 ekovo
4020 efdyc	**4070** efodi	**4120** egcus	**4170** egmyg	**4220** ehadu	**4270** ehlyb	**4320** ejasy	**4370** ejrel	**4420** ekduv	**4470** ekoxe
4021 efeid	4071 efoho	4121 egdab	4171 egnal	4221 ehafa	4271 ehmaj	4321 ejaxt	4371 ejruw	4421 ekect	4471 ekpav
4022 efelz	4072 efonu	4122 egdev	4172 egnek	4222 ehale	4272 ehmeh	4322 ejbah	4372 ejske	4422 ekegu	4472 ekpex
4023 efemy	4073 efost	4123 egdic	4173 egnyo	4223 ehani	4273 ehmik	4323 ejbey	4373 ejsli	4423 ekeko	4473 ekpli
4024 efept	4074 efoxy	4124 egdon	4174 egobs	4224 ehaph	4274 ehmop	4324 ejbot	4374 ejspu	4424 ekems	4474 ekpop
4025 efere	4075 efpag	4125 egdry	4175 egoda	4225 eharx	4275 ehmur	4325 ejcid	4375 ejtaf	4425 ekepy	4475 ekpur
4026 efeso	4076 efpeb	4126 egduk	4176 egofi	4226 ehato	4276 ehmyd	4326 ejcos	4376 ejtes	4426 ekexa	4476 ekpyk
4027 efevu	4077 efpim	4127 egebu	4177 egojo	4227 ehbak	4277 ehnax	4327 ejdat	4377 ejtic	4427 ekezi	4477 ekraz
4028 effav	4078 efpow	4128 egefa	4178 egoku	4228 ehbem	4278 ehner	4328 ejdru	4378 ejtog	4428 ekfah	4478 ekrep
4029 effey	4079 efpun	4129 egens	4179 egowl	4229 ehbid	4279 ehnif	4329 ejean	4379 ejtym	4429 ekfem	4479 ekrif
4030 effiz	**4080** efraw	**4130** egeph	**4180** egpam	**4230** ehbob	**4280** ehnup	**4330** ejefu	**4380** ejuch	**4430** ekfyl	**4480** ekrok
4031 effop	4081 efref	4131 egerd	4181 egpel	4231 ehbuf	4281 ehnyg	4331 ejerk	4381 ejudu	4431 ekgal	4481 ekryc
4032 effur	4082 efroz	4132 egeto	4182 egpiv	4232 ehbyx	4282 ehoch	4332 ejfal	4382 ejums	4432 ekgen	4482 eksab
4033 efgax	4083 efsah	4133 egewi	4183 egpot	4233 ehcaw	4283 ehode	4333 ejfet	4383 ejunk	4433 ekgiv	4483 eksev
4034 efger	4084 efsem	4134 egfac	4184 egpud	4234 ehcoy	4284 ehoja	4334 ejfif	4384 ejuri	4434 ekgos	4484 eksis
4035 efgib	4085 efsip	4135 egfeg	4185 egran	4235 ehcut	4285 ehomi	4335 ejfoc	4385 ejuto	4435 ekgut	4485 ekson
4036 efgov	4086 efsob	4136 egfib	4186 egrew	4236 ehdac	4286 ehops	4336 ejgeh	4386 ejuve	4436 ekheh	4486 eksup
4037 efgup	4087 efsuf	4137 egfom	4187 egrik	4237 ehdeg	4287 ehorf	4337 ejgow	4387 ejuxa	4437 ekhin	4487 eksym
4038 efhaj	4088 eftam	4138 egfru	4188 egrod	4238 ehdin	4288 ehotu	4338 ejgra	4388 ejvex	4438 ekhuy	4488 ektap
4039 efhir	4089 eftep	4139 eggec	4189 egrum	4239 ehdul	4289 ehpaz	4339 ejhag	4389 ejvig	4439 ekidu	4489 ektub
4040 efhox	**4090** eftiv	**4140** eggif	**4190** egryl	**4240** ehdyp	**4290** ehpep	**4340** ejhem	**4390** ejvyn	**4440** ekily	[illegible]
4041 efhuk	4091 eftot	4141 eggol	4191 egsap	4241 ehefi	4291 ehpok	4341 ejhyb	4391 ejwab	4441 ekino	4491 ekubi
4042 efics	4092 eftul	4142 eggub	4192 egses	4242 ehega	4292 ehpus	4342 ejiky	4392 ejwen	4442 ekiva	4492 ekulo
4043 efigo	4093 efudy	4143 eghav	4193 egsiz	4243 eheke	4293 ehrab	4343 ejilu	4393 ejwil	4443 ekise	4493 ekups
4044 efika	4094 efval	4144 eghit	4194 egsof	4244 ehelm	4294 ehrev	4344 ejins	4394 ejwox	4444 ekjel	4494 ekuxu
4045 efiph	4095 efven	4145 eghoy	4195 egsuc	4245 ehepu	4295 ehron	4345 ejipa	4395 ejxas	4445 ekjig	4495 ekuza
4046 efity	4096 efvit	4146 eghul	4196 egsyr	4246 eheth	4296 ehruk	4346 ejkik	4396 ejxer	4446 ekjus	4496 ekvac
4047 efizu	4097 efvos	4147 egibe	4197 egtaw	4247 ehfig	4297 ehsad	4347 ejkof	4397 ekafo	4447 ekkan	4497 ekveg
4048 efjat	4098 efway	4148 egica	4198 egtex	4248 ehgan	4298 ehsec	4348 ejkux	4398 ekahy	4448 ekkey	4498 ekvom
4049 efjew	4099 efwex	4149 egifo	4199 egthi	4249 ehgew	4299 ehsub	4349 ejlak	4399 ekaix	4449 ekkip	4499 ekwak

4500 [illegible]	4550 elleh	4600 elyet	4650 emipu	4700 emtuk	4750 enhib	4800 ensun	4850 epgaf	4900 epseg	4950 eqexi
4501 [illegible]	4551 ellit	4601 elyup	4651 emiti	4701 emufe	4751 enhov	4801 ensyd	4851 epged	4901 epsuk	4951 eqeze
4502 [illegible]	4552 ellur	4602 emabe	4652 emiwa	4702 emuku	4752 enhuf	4802 entah	4852 epgno	4902 epsyn	4952 eqfay
4503 [illegible]	4553 ellyd	4603 emack	4653 emjah	4703 emuni	4753 eniba	4803 entob	4853 epgry	4903 eptak	4953 eqfex
4504 [illegible]	4554 elmer	4604 emaft	4654 emjer	4704 emvak	4754 enife	4804 entyn	4854 epguz	4904 eptet	4954 eqfil
4505 elada	4555 elmov	4605 emagi	4655 emjis	4705 emvem	4755 enimi	4805 enuco	4855 ephax	4905 eptij	4955 eqfoh
4506 elaik	4556 elnaz	4606 emahu	4656 emkaz	4706 emvic	4756 eniru	4806 enuky	4856 ephic	4906 eptoc	4956 eqfuz
4507 elans	4557 elnep	4607 emaja	4657 emkep	4707 emvul	4757 enjas	4807 enume	4857 ephub	4907 eptun	4957 eqgas
4508 elarg	4558 elnim	4608 emamy	4658 emkin	4708 emweh	4758 enjet	4808 enuph	4858 epick	4908 eptyd	4958 eqgek
4509 elavu	4559 elnok	4609 emarn	4659 emkok	4709 emyle	4759 enjor	4809 enuvi	4859 epidy	4909 epugi	4959 eqgit
4510 elbal	4560 elnus	4610 emasp	4660 emkus	4710 emyot	4760 enkav	4810 enuzu	4860 epifi	4910 epupo	4960 eqgor
4511 elbos	4561 elofo	4611 emawo	4661 emkym	4711 emyub	4761 enkew	4811 envay	4861 epips	4911 epurb	4961 eqgug
4512 elcab	4562 eloke	4612 emban	4662 emlal	4712 emzam	4762 enkiz	4812 envef	4862 epire	4912 epusa	4962 eqheo
4513 eleli	4563 elold	4613 embet	4663 emlex	4713 emzel	4763 enkry	4813 envid	4863 epita	4913 epute	4963 eqhiv
4514 elcon	4564 elont	4614 embiz	4664 emlib	4714 emzoy	4764 enlax	4814 envoz	4864 epjul	4914 epveh	4964 eqids
4515 eldam	4565 elopu	4615 embod	4665 emloh	4715 enaci	4765 enler	4815 enwaw	4865 epker	4915 epvip	4965 eqiga
4516 eldel	4566 elorb	4616 embum	4666 emlut	4716 enado	4766 enlot	4816 enwem	4866 epkig	4916 epvow	4966 eqime
4517 eldro	4567 elovi	4617 embyc	4667 emlyr	4717 enafu	4767 enlup	4817 enwis	4867 epkup	4917 epvus	4967 eqinu
4518 eldud	4568 eloxa	4618 emcap	4668 emmac	4718 enaga	4768 enmaz	4818 enzug	4868 epkyf	4918 epvyt	4968 eqiri
4519 eldyt	4569 elpac	4619 emces	4669 emmeg	4719 enalk	4769 enmep	4819 epafy	4869 eplem	4919 epwav	4969 eqixo
4520 elebe	4570 elpeg	4620 emciv	4670 emmif	4720 enash	4770 enmin	4820 epals	4870 eplin	4920 epwec	4970 eqkef
4521 elecs	4571 elpin	4621 emcof	4671 emmom	4721 enbam	4771 enmok	4821 epamu	4871 eplob	4921 epwon	4971 eqkoz
4522 eleju	4572 elplu	4622 emcuc	4672 emnew	4722 enbel	4772 enmus	4822 epank	4872 epluf	4922 epxol	4972 eqkru
4523 elerf	4573 elpom	4623 emcyx	4673 emnik	4723 enbiv	4773 ennab	4823 epapi	4873 eplyx	4923 epzid	4973 eqlah
4524 eleta	4574 elpyx	4624 emdar	4674 emnol	4724 enboy	4774 ennev	4824 epavo	4874 epmal	4924 epzle	4974 eqlen
4525 elewo	4575 elrad	4625 emdez	4675 emobo	4725 enbud	4775 ennig	4825 epawa	4875 epmen	4925 epzot	4975 eqlos
4526 elfan	4576 elrec	4626 emdip	4676 emofy	4726 encac	4776 ennon	4826 epaxe	4876 epmos	4926 eqacu	4976 eqluw
4527 elfew	4577 elrol	4627 emdog	4677 emome	4727 enceg	4777 ennuk	4827 epbaj	4877 epmyr	4927 eqams	4977 eqmam
4528 elfod	4578 elrub	4628 emdux	4678 emonc	4728 enclu	4778 ennyp	4828 epbil	4878 epnam	4928 eqata	4978 eqmel
4529 elfri	4579 elsaf	4629 emeat	4679 emopa	4729 encom	4779 enody	4829 epboz	4879 epnel	4929 eqave	4979 eqmot
4530 elfum	4580 elsed	4630 emedy	4680 emots	4730 encys	4780 enola	4830 epbys	4880 epnit	4930 eqaxy	4980 eqmud
4531 elfys	4581 elshy	4631 emefo	4681 empaf	4731 endey	4781 enomo	4831 epcan	4881 epnud	4931 eqazi	4981 eqnar
4532 elgap	4582 elsig	4632 emelv	4682 emped	4732 endif	4782 enong	4832 epcey	4882 epnyc	4932 eqbab	4982 eqnez
4533 elges	4583 elsoc	4633 ememu	4683 empoc	4733 endod	4783 enose	4833 epcod	4883 epocu	4933 eqbev	4983 eqnip
4534 elgiz	4584 elsuz	4634 emerg	4684 empuz	4734 endut	4784 enour	4834 epcum	4884 epoko	4934 eqbis	4984 eqnog
4535 elgof	4585 eltav	4635 emese	4685 emrag	4735 enecu	4785 enpad	4835 epdaw	4885 epoma	4935 eqbon	4985 eqnux
4536 elgym	4586 eltem	4636 emfas	4686 emreb	4736 enego	4786 enpec	4836 epdex	4886 epope	4936 eqbuk	4986 eqoco
4537 elhak	4587 eltis	4637 emfix	4687 emrid	4737 eneja	4787 enpij	4837 epdis	4887 eposy	4937 eqcem	4987 eqoki
4538 elhey	4588 eltuf	4638 emfug	4688 emrox	4738 eneli	4788 enpol	4838 epdly	4888 epoti	4938 eqciz	4988 eqolt
4539 elhip	4589 elubs	4639 emgav	4689 emrun	4739 enerx	4789 enpub	4839 epdur	4889 eppas	4939 eqcob	4989 eqons
4540 elhux	4590 eluck	4640 emgim	4690 emryf	4740 eneve	4790 enpyr	4840 epejo	4890 eppek	4940 eqcra	4990 eqowe
4541 elich	4591 eluly	4641 emgop	4691 emsaw	4741 enews	4791 enraf	4841 epelf	4891 eppix	4941 eqdap	4991 eqpan
4542 eljez	4592 eluno	4642 emgur	4692 emsef	4742 enfap	4792 enred	4842 epene	4892 eppor	4942 eqdes	4992 eqpet
4543 eljog	4593 elura	4643 emgyn	4693 emsit	4743 enfit	4793 enril	4843 epeva	4893 eppug	4943 eqdof	4993 eqpib
4544 elkas	4594 eluzi	4644 emhab	4694 emsoz	4744 enfro	4794 enroc	4844 epexu	4894 eprap	4944 eqduc	4994 eqpod
4545 elkek	4595 elvag	4645 emhek	4695 emsuv	4745 enful	4795 enruz	4845 epfar	4895 epres	4945 eqelu	4995 eqpum
4546 elkid	4596 elvox	4646 emhig	4696 emsys	4746 engar	4796 ensag	4846 epfez	4896 eprof	4946 eqepo	4996 eqrac
4547 elkug	4597 elvun	4647 emhon	4697 emten	4747 engez	4797 enseb	4847 epfik	4897 epruc	4947 eqeha	4997 eqreg
4548 elkyp	4598 elwir	4648 emhup	4698 emtha	4748 engog	4798 ensik	4848 epfog	4898 epryb	4948 eqeny	4998 eqrom
4549 ellaw	4599 elwob	4649 eminy	4699 emtry	4749 enhat	4799 ensox	4849 epfux	4899 epsac	4949 eqets	4999 eqsat

eqsoy eqsul eqtax eqter

KEY ID:A10154

SECTION A

5000 eqtid	**5050** erjay	**5100** erwig	**5150** esjen	**5200** esvej	**5250** etkyx	**5300** evacs	**5350** evith	**5400** evumy	**5450** ewjad
5001 eqtup	5051 erjef	5101 erwom	5151 esjos	5201 esvib	5251 etlaz	5301 evald	5351 evjak	5401 evunz	5451 ewjek
5002 equfu	5052 erjid	5102 erwyn	5152 esjut	5202 esvor	5252 etlic	5302 evapy	5352 evjem	5402 evver	5452 ewjow
5003 equjo	5053 erjot	5103 eryem	5153 eskam	5203 eswah	5253 etlus	5303 evarm	5353 evjob	5403 evvov	5453 ewjur
5004 equlk	5054 erjup	5104 eryux	5154 eskel	5204 eswey	5254 etlyl	5304 evauk	5354 evjuf	5404 evvup	5454 ewkaw
5005 eqvaz	5055 erkak	5105 erzaf	5155 eskif	5205 eswho	5255 etmay	5305 evawn	5355 evkal	5405 evwaz	5455 ewkem
5006 eqvey	5056 erkeh	5106 erzed	5156 eskro	5206 esyra	5256 etmis	5306 evaxi	5356 evken	5406 evwet	5456 ewkot
5007 eqvim	5057 erkob	5107 erzib	5157 eskud	5207 eszef	5257 etmul	5307 evaza	5357 evkos	5407 evwod	5457 ewkuk
5008 eqvok	5058 erkuf	5108 erzly	5158 eslan	5208 etaec	5258 etnex	5308 evbac	5358 evkyk	5408 evyew	5458 ewlav
5009 eqvuf	5059 erkys	5109 erzoc	5159 eslet	5209 etaha	5259 etnil	5309 evbeg	5359 evlam	5409 evyla	5459 ewlep
5010 eqwag	**5060** erlat	**5110** erzuz	**5160** eslis	**5210** etain	**5260** etnoy	**5310** evbif	**5360** evlel	**5410** evyus	**5460** ewlok
5011 eqweb	5061 erlov	5111 esagy	5161 eslod	5211 etalp	5261 etnub	5311 evblu	5361 evlix	5411 evzah	5461 ewlyc
5012 eqwin	5062 erlul	5112 esano	5162 eslum	5212 etamb	5262 etofa	5312 evbro	5362 evlud	5412 evzin	5462 ewmat
5013 eraku	5063 erlyg	5113 esare	5163 esmap	5213 etark	5263 etoge	5313 evcad	5363 evman	5413 ewabs	5463 ewmey
5014 erany	5064 ermab	5114 esasi	5164 esmes	5214 etayo	5264 etolm	5314 evceo	5364 evmex	5414 ewajo	5464 ewmod
5015 erapt	5065 ermev	5115 esatu	5165 esmiz	5215 etboj	5265 etopo	5315 evcri	5365 evmig	5415 ewake	5465 ewmup
5016 erars	5066 ermim	5116 esawl	5166 esmof	5216 etcat	5266 etorn	5316 evcub	5366 evmoh	5416 ewalu	5466 ewnak
5017 erbaz	5067 ermon	5117 esbik	5167 esmuc	5217 etcen	5267 etosk	5317 evcyn	5367 evmum	5417 ewarc	5467 ewnec
5018 erbok	5068 ermuk	5118 esbol	5168 esmyt	5218 etclo	5268 etozu	5318 evdaf	5368 evmyl	5418 ewavi	5468 ewniz
5019 erbre	5069 ernan	5119 escaf	5169 esnac	5219 etcuf	5269 etpef	5319 evded	5369 evnap	5419 ewbex	5469 ewnof
5020 erbus	**5070** erney	**5120** esced	**5170** esneg	**5220** etcyr	**5270** etply	**5320** evdil	**5370** evnes	**5420** ewcah	**5470** ewnut
5021 erbyt	5071 ernod	5121 escoc	5171 esniv	5221 etdew	5271 etpru	5321 evdoc	5371 evnic	5421 ewcig	5471 ewoby
5022 ercam	5072 ernum	5122 escuz	5172 esnom	5222 etdir	5272 etraj	5322 evduz	5372 evnot	5422 ewcul	5472 ewocs
5023 ercel	5073 eromy	5123 escyk	5173 esnys	5223 etdot	5273 etrey	5323 evdyx	5373 evobu	5423 ewcym	5473 ewogn
5024 ercow	5074 erono	5124 esdag	5174 esodo	5224 etdum	5274 etrix	5324 eveba	5374 evogy	5424 ewdet	5474 ewoka
5025 ercud	5075 erork	5125 esdeb	5175 esoli	5225 etdys	5275 etser	5325 evech	5375 evols	5425 ewdom	5475 ewole
5026 ercyp	5076 erosa	5126 esdid	5176 esoms	5226 eteal	5276 etsid	5326 evedo	5376 evond	5426 ewdri	5476 eworo
5027 erdal	5077 erovu	5127 esdox	5177 esonz	5227 eteby	5277 etsky	5327 evege	5377 evorc	5427 ewdyl	5477 ewotz
5028 erdik	5078 erpyd	5128 esdun	5178 esory	5228 eteco	5278 etsuh	5328 eveki	5378 evoto	5428 ewebi	5478 ewoxi
5029 erdos	5079 errit	5129 eseky	5179 esowa	5229 eteig	5279 ettar	5329 evemp	5379 evoze	5429 eweft	5479 ewpaj
5030 erduy	**5080** erryx	**5130** eseld	**5180** esoxu	**5230** eters	**5280** ettel	**5330** evesu	**5380** evpar	**5430** ewelk	**5480** ewpho
5031 ereka	5081 ersek	5131 esepi	5181 espax	5231 eteux	5281 ettif	5331 evfag	5381 evpez	5431 ewena	5481 ewpuy
5032 erelb	5082 ersor	5132 eseru	5182 espov	5232 etfli	5282 ettom	5332 evfeb	5382 evpik	5432 ewerp	5482 ewray
5033 erenu	5083 erspi	5133 esete	5183 espup	5233 etfos	5283 ettug	5333 evfox	5383 evpog	5433 ewesh	5483 ewric
5034 eresi	5084 ersug	5134 esfat	5184 esrav	5234 etfre	5284 ettyb	5334 evfun	5384 evpyt	5434 eweuc	5484 ewroh
5035 erflo	5085 ersyl	5135 esfim	5185 esrex	5235 etglu	5285 etufy	5335 evgat	5385 evras	5435 ewfol	5485 ewsib
5036 erfry	5086 ertac	5136 esfoz	5186 esrul	5236 ethog	5286 etula	5336 evgef	5386 evrek	5436 ewfra	5486 ewsko
5037 ergag	5087 ertil	5137 esgak	5187 esrym	5237 ethud	5287 etumo	5337 evgid	5387 evrib	5437 ewfuf	5487 ewstu
5038 ergir	5088 ertof	5138 esgem	5188 essar	5238 etiap	5288 etupi	5338 evgoz	5388 evror	5438 ewgaz	5488 ewswe
5039 ergle	5089 ertye	5139 esgly	5189 essez	5239 etibo	5289 etuvu	5339 evgul	5389 evrug	5439 ewgli	5489 ewsyf
5040 ergun	**5090** erugo	**5140** esgob	**5190** essin	**5240** etile	**5290** etvep	**5340** evgyb	**5390** evsim	**5440** ewgoc	**5490** ewtan
5041 erhah	5091 eruli	5141 esguf	5191 essog	5241 etink	5291 etwad	5341 evhaw	5391 evsly	5441 ewgre	5491 ewtop
5042 erheg	5092 erumu	5142 eshow	5192 essux	5242 etixu	5292 etweg	5342 evhok	5392 evsop	5442 ewgud	5492 ewtux
5043 eriby	5093 erush	5143 esibu	5193 estaz	5243 etjip	5293 etwhy	5343 evhut	5393 evthy	5443 ewhez	5493 ewual
5044 erico	5094 eruta	5144 esify	5194 estix	5244 etjok	5294 etwim	5344 evhyc	5394 evtit	5444 ewhid	5494 ewubu
5045 erifa	5095 ervav	5145 esige	5195 estyr	5245 etjun	5295 etwov	5345 evibi	5395 evtow	5445 ewhob	5495 ewude
5046 erils	5096 ervet	5146 esind	5196 esudi	5246 etkah	5296 etzem	5346 evicu	5396 evtre	5446 ewimy	5496 ewufi
5047 eritu	5097 ervif	5147 esisl	5197 esugu	5247 etket	5297 etziz	5347 eviko	5397 evube	5447 ewini	5497 ewvis
5048 erive	5098 ervoy	5148 esivo	5198 esuko	5248 etkow	5298 etzob	5348 eving	5398 evuja	5448 ewisk	5498 ewvon
5049 erizi	5099 erwax	5149 esjal	5199 esvas	5249 etkri	5299 evabo	5349 evisy	5399 evult	5449 ewizo	5499 ewwaf

5500 ewwel	5550 exiha	5600 eybra	5650 ezdyn	5700 ezpyc	5750 fafon	5800 fease	5850 feojo	5900 fieke	5950 firev
5501 ewwip	5551 exims	5601 eycyt	5651 ezebo	5701 ezrax	5751 fagac	5801 febax	5851 feoku	5901 fielm	5951 firiz
5502 ewwog	5552 exirp	5602 eydow	5652 ezede	5702 ezrhe	5752 fageg	5802 feber	5852 fepiv	5902 fiepu	5952 firuk
5503 ewxar	5553 exitz	5603 eydre	5653 ezeku	5703 ezrov	5753 fagin	5803 feblo	5853 fepud	5903 fieth	5953 fisad
5504 ewxen	5554 exjav	5604 eyelo	5654 ezela	5704 ezrup	5754 fagom	5804 febup	5854 ferew	5904 fifam	5954 fispy
5505 ewxim	5555 exjib	5605 eyerb	5655 ezern	5705 ezsaz	5755 fagyl	5805 fecaz	5855 ferik	5905 fifel	5955 fited
5506 ewyas	5556 exjod	5606 eyetz	5656 ezfaw	5706 ezsep	5756 fahad	5806 fecro	5856 feryl	5906 fifig	5956 fitip
5507 ewyef	5557 exjum	5607 eyfuc	5657 ezfen	5707 ezsmi	5757 fahif	5807 fecus	5857 feses	5907 fifot	5957 fituc
5508 ewygo	5558 exkay	5608 eygah	5658 ezfic	5708 ezsus	5758 fahyr	5808 fedev	5858 fesiz	5908 fifud	5958 fityt
5509 ewyne	5559 exkex	5609 eyhef	5659 ezfut	5709 eztal	5759 faigi	5809 fedic	5859 fesof	5909 figan	5959 fiubo
5510 ewyum	5560 exklo	5610 eykom	5660 ezgip	5710 ezteg	5760 fainz	5810 feduk	5860 fesyr	5910 figew	5960 fiuka
5511 ewzap	5561 exkut	5611 eyofe	5661 ezgru	5711 eztin	5761 fairo	5811 feebu	5861 fetex	5911 figic	5961 fiulu
5512 ewzif	5562 exlab	5612 eyohn	5662 ezhaf	5712 eztos	5762 fajeh	5812 feele	5862 fethi	5912 figod	5962 fiune
5513 ewzus	5563 exley	5613 eyoty	5663 ezhuz	5713 eztum	5763 fakaf	5813 feens	5863 fetuy	5913 figum	5963 fivat
5514 exaep	5564 exlil	5614 eypri	5664 ezidi	5714 eztyx	5764 faked	5814 feerd	5864 feuga	5914 fihay	5964 fivix
5515 exait	5565 exlop	5615 eyshe	5665 ezifu	5715 ezucu	5765 fakli	5815 feeph	5865 feumi	5915 fihor	5965 fivoc
5516 exand	5566 exluk	5616 eysny	5666 ezilk	5716 ezuki	5766 fakoc	5816 feewi	5866 feutu	5916 fihun	5966 fiwal
5517 exary	5567 exmew	5617 eysok	5667 ezisa	5717 ezung	5767 faliz	5817 fefeg	5867 fevar	5917 fihym	5967 fiwiv
5518 exaus	5568 exmoy	5618 eytay	5668 ezite	5718 ezupa	5768 falox	5818 fefom	5868 fevez	5918 fijap	5968 fixis
5519 exbat	5569 exnyt	5619 eytle	5669 ezivy	5719 ezuso	5769 falyt	5819 fefru	5869 fevog	5919 fijes	5969 fizag
5520 exbeb	5570 exoen	5620 eytro	5670 ezkap	5720 ezvad	5770 famah	5820 fegec	5870 fewid	5920 fijil	5970 foaco
5521 exbir	5571 exoid	5621 eyuel	5671 ezkes	5721 ezvec	5771 famef	5821 fegif	5871 feyah	5921 fijof	5971 foada
5522 exbow	5572 exows	5622 eyuha	5672 ezkim	5722 ezvir	5772 famic	5822 fegol	5872 feyen	5922 fijux	5972 foame
5523 exbul	5573 exozo	5623 eyury	5673 ezkuc	5723 ezvol	5773 famoz	5823 fegub	5873 feyow	5923 fikar	5973 foans
5524 excal	5574 expip	5624 eyvop	5674 ezlar	5724 ezwac	5774 fanid	5824 fehav	5874 fezet	5924 fikib	5974 foarg
5525 excef	5575 expla	5625 eywas	5675 ezlez	5725 ezwed	5775 faofu	5825 fehul	5875 fezop	5925 fikog	5975 foavu
5526 excim	5576 expof	5626 eywep	5676 ezlid	5726 ezzat	5776 faoly	5826 feica	5876 fezur	5926 filas	5976 fobal
5527 excot	5577 expre	5627 eywib	5677 ezlog	5727 ezzom	5777 faomp	5827 feifo	5877 fiacy	5927 filek	5977 foben
5528 exdad	5578 exrah	5628 ezaby	5678 ezlux	5728 faabi	5778 faort	5828 feigu	5878 fiadu	5928 filow	5978 focab
5529 exdec	5579 exrog	5629 ezadz	5679 ezlym	5729 faady	5779 faosi	5829 feits	5879 fiafa	5929 filug	5979 focli
5530 exdol	5580 exsak	5630 ezaho	5680 ezmas	5730 faama	5780 faowo	5830 fejaf	5880 fiani	5930 filyb	5980 focon
5531 exdub	5581 exshu	5631 ezalc	5681 ezmek	5731 faapo	5781 farez	5831 fejed	5881 fiaph	5931 fimaj	5981 focre
5532 exdym	5582 exsme	5632 ezana	5682 ezmit	5732 fabav	5782 faryn	5832 fejoc	5882 fiarx	5932 fimeh	5982 fodam
5533 exeda	5583 extek	5633 ezari	5683 ezmor	5733 fabij	5783 fatas	5833 fekag	5883 fibak	5933 fimop	5983 fodiv
5534 exemb	5584 extho	5634 ezasm	5684 ezmug	5734 faboh	5784 fatey	5834 fekeb	5884 fibem	5934 fimur	5984 fodro
5535 exeni	5585 extuz	5635 ezaxu	5685 eznav	5735 facib	5785 faufa	5835 fekil	5885 fibid	5935 fimyd	5985 fodud
5536 exeps	5586 exupu	5636 ezbag	5686 ezneh	5736 fadaz	5786 faunc	5836 fekun	5886 fibob	5936 finif	5986 fodyt
5537 exerz	5587 exusi	5637 ezbef	5687 eznur	5737 fadep	5787 faurg	5837 felef	5887 fibuf	5937 finov	5987 foebe
5538 exetu	5588 exvur	5638 ezbib	5688 ezobe	5738 fadok	5788 fauxo	5838 feloz	5888 fibyx	5938 finyg	5988 foecs
5539 exeye	5589 exwam	5639 ezbox	5689 ezock	5739 faedi	5789 fauze	5839 feluv	5889 ficaw	5939 fioch	5989 foeju
5540 exfed	5590 exwer	5640 ezbun	5690 ezolo	5740 faemo	5790 faves	5840 femak	5890 ficir	5940 fiode	5990 foend
5541 exfis	5591 exwiz	5641 ezbyl	5691 ezomu	5741 faeng	5791 faviv	5841 femem	5891 ficoy	5941 fioja	5991 foerf
5542 exgaw	5592 exyap	5642 ezcay	5692 ezoni	5742 faerl	5792 fawor	5842 femip	5892 ficut	5942 fiomi	5992 foeta
5543 exget	5593 exyel	5643 ezcet	5693 ezopy	5743 faesy	5793 faxel	5843 femuf	5893 fidac	5943 fiops	5993 foewo
5544 exgig	5594 exzas	5644 ezcoz	5694 ezota	5744 faewa	5794 fayog	5844 femyg	5894 fideg	5944 fiorf	5994 fofan
5545 exgro	5595 exzez	5645 ezdak	5695 ezpab	5745 faexe	5795 fazek	5845 fenal	5895 fidin	5945 fiotu	5995 fofew
5546 exguy	5596 exzor	5646 ezdem	5696 ezpev	5746 fafab	5796 fazir	5846 fenek	5896 fidul	5946 fipaz	5996 fofod
5547 exhan	5597 exzuc	5647 ezdig	5697 ezpif	5747 fafev	5797 feaba	5847 fenyo	5897 fidyp	5947 fipep	5997 fofri
5548 exhoc	5598 eyaig	5648 ezdob	5698 ezpon	5748 fafip	5798 feaju	5848 feobs	5898 fiefi	5948 fipit	5998 fofum
5549 exhyl	5599 eyalt	5649 ezduf	5699 ezpuk	5749 faflu	5799 fealy	5849 feoda	5899 fiega	5949 fipok	5999 fofys

fogap foges fogiz fogof

[redacted]

SECTION A

6000 foguc	**6050** forub	**6100** fuegu	**6150** fupop	**6200** fyets	**6250** fypum	**6300** gaewe	**6350** gaown	**6400** gedif	**6450** gepol
6001 fogym	6051 fosaf	6101 fueko	6151 fupur	6201 fyexi	6251 fyrus	6301 gafaz	6351 gaoxo	6401 gedod	6451 gepub
6002 fohak	6052 fosed	6102 fuems	6152 fupyk	6202 fyeze	6252 fysat	6302 gafep	6352 gapak	6402 geecu	6452 gepyr
6003 fohey	6053 foshy	6103 fuepy	6153 furaz	6203 fyfay	6253 fysoy	6303 gafin	6353 gapem	6403 geego	6453 geraf
6004 fohip	6054 fosig	6104 fuexa	6154 furep	6204 fyfex	6254 fysul	6304 gafly	6354 gapid	6404 geeja	6454 gered
6005 fohoz	6055 fosoc	6105 fuezi	6155 furif	6205 fyfil	6255 fytax	6305 gafok	6355 gaplo	6405 geeli	6455 geril
6006 fohux	6056 fosuz	6106 fufah	6156 furok	6206 fyfoh	6256 fyter	6306 gafus	6356 gapry	6406 geeve	6456 geroc
6007 foido	6057 fotav	6107 fufem	6157 furud	6207 fyfuz	6257 fytid	6307 gagev	6357 gapuf	6407 gefap	6457 geruz
6008 foine	6058 fotem	6108 fufyl	6158 furyc	6208 fygas	6258 fytup	6308 gagis	6358 garal	6408 gefit	6458 gesag
6009 foisi	6059 fotis	6109 fugal	6159 fusev	6209 fygek	6259 fytys	6309 gagon	6359 garen	6409 gefro	6459 geseb
6010 foiza	**6060** fotuf	**6110** fugiv	**6160** fusis	**6210** fygit	**6260** fyufu	**6310** gagyr	**6360** garij	**6410** geful	**6460** gesik
6011 fojar	6061 foubs	6111 fugut	6161 fuson	6211 fygor	6261 fyujo	6311 gahap	6361 garut	6411 gegar	6461 gesox
6012 fojez	6062 fouck	6112 fuheh	6162 fusym	6212 fygug	6262 fyulk	6312 gahew	6362 gasif	6412 gegez	6462 gesun
6013 fojog	6063 fouly	6113 fuhin	6163 futec	6213 fyhec	6263 fyvaz	6313 gahof	6363 gasyk	6413 gegog	6463 gesyd
6014 fokas	6064 founo	6114 fuhuy	6164 futir	6214 fyhiv	6264 fyvey	6314 gahuc	6364 gatat	6414 gehat	6464 getyn
6015 fokek	6065 foura	6115 fuidu	6165 futub	6215 fyids	6265 fyvim	6315 gahyd	6365 gatez	6415 gehib	6465 geuco
6016 fokid	6066 fouzi	6116 fuily	6166 futyp	6216 fyiga	6266 fyvok	6316 gaibs	6366 gatig	6416 gehov	6466 geuda
6017 fokor	6067 fovag	6117 fuino	6167 fuvac	6217 fyime	6267 fyvuf	6317 gaice	6367 gatoy	6417 gehuf	6467 geuky
6018 fokug	6068 fovil	6118 fuiva	6168 fuveg	6218 fyinu	6268 fywag	6318 gaipi	6368 gatru	6418 geiba	6468 geuvi
6019 fokyp	6069 fovox	6119 fuize	6169 fuvom	6219 fyiri	6269 fyweb	6319 gairl	6369 gaufo	6419 geife	6469 geuzu
6020 folaw	**6070** fovun	**6120** fujel	**6170** fuwak	**6220** fyixo	**6270** fywin	**6320** gajac	**6370** gauma	**6420** geimi	**6470** gevay
6021 foleh	6071 fowir	6121 fujig	6171 fuwic	6221 fykef	6271 fywow	6321 gajeg	6371 gaund	6421 geiru	6471 gevef
6022 folit	6072 fowob	6122 fujus	6172 fuwol	6222 fykoz	6272 fywry	6322 gajuw	6372 gause	6422 geiso	6472 gevid
6023 folur	6073 foyet	6123 fukan	6173 fuzax	6223 fykru	6273 fyxal	6323 gakad	6373 gauti	6423 gejas	6473 gevoz
6024 folyd	6074 foyup	6124 fukey	6174 fuzer	6224 fylah	6274 fyxur	6324 gakec	6374 gavan	6424 gejet	6474 gewaw
6025 fomax	6075 fozic	6125 fukip	6175 fyacu	6225 fylen	6275 gaabu	6325 gakir	6375 gavik	6425 gejor	6475 gewem
6026 fomer	6076 fozuk	6126 fukod	6176 fyams	6226 fylos	6276 gaaca	6326 gakol	6376 gavum	6426 gekav	6476 gewis
6027 fomib	6077 fuafo	6127 fukum	6177 fyaro	6227 fyluw	6277 gaads	6327 gakub	6377 gawar	6427 gekew	6477 geyak
6028 fomov	6078 fuahy	6128 fulap	6178 fyata	6228 fymam	6278 gaafi	6328 galaf	6378 gawes	6428 gekiz	6478 gezal
6029 fomut	6079 fuang	6129 fules	6179 fyave	6229 fymel	6279 gaago	6329 galed	6379 gawot	6429 gekry	6479 gezen
6030 fomyn	**6080** fuape	**6130** fulik	**6180** fyaxy	**6230** fymir	**6280** gaalm	**6330** galip	**6380** gayaw	**6430** gelax	**6480** gezip
6031 fonaz	6081 fuara	6131 fulof	6181 fyazi	6231 fymot	6281 gaane	6331 galoc	6381 gazey	6431 geler	6481 gezos
6032 fonep	6082 fubaf	6132 fuluc	6182 fybab	6232 fymud	6282 gaaty	6332 galuz	6382 gazog	6432 gelot	6482 gezug
6033 fonok	6083 fubed	6133 fumez	6183 fybev	6233 fynar	6283 gaawk	6333 galyn	6383 gazul	6433 gelup	6483 giame
6034 fonus	6084 fubit	6134 fumog	6184 fybis	6234 fynez	6284 gabas	6334 gamag	6384 geaci	6434 gemaz	6484 gians
6035 foofo	6085 fuboc	6135 fumux	6185 fybon	6235 fynip	6285 gabek	6335 gameb	6385 geado	6435 gemep	6485 giarg
6036 fooke	6086 fubry	6136 fumys	6186 fybuk	6236 fynog	6286 gabli	6336 gamil	6386 geafu	6436 gemin	6486 giavu
6037 foold	6087 fubuz	6137 funas	6187 fycem	6237 fynux	6287 gabor	6337 gamox	6387 geaga	6437 genab	6487 gibal
6038 foont	6088 fucag	6138 funet	6188 fyciz	6238 fynyk	6288 gabug	6338 gamun	6388 gealk	6438 genev	6488 gibix
6039 foorb	6089 fuceb	6139 funor	6189 fycob	6239 fyoco	6289 gacav	6339 gamyc	6389 geash	6439 genig	6489 gicli
6040 foovi	**6090** fucil	**6140** funug	**6190** fycra	**6240** fyohm	**6290** gacit	**6340** ganah	**6390** gebam	**6440** genon	**6490** gicre
6041 fooxa	6091 fucox	6141 funyx	6191 fycyc	6241 fyoki	6291 gacop	6341 ganef	6391 gebel	6441 genuk	6491 gidam
6042 fopac	6092 fucun	6142 fuoga	6192 fydap	6242 fyolt	6292 gacur	6342 ganib	6392 gebiv	6442 genyp	6492 gidel
6043 fopeg	6093 fucyd	6143 fuolu	6193 fydes	6243 fyons	6293 gacyg	6343 ganoz	6393 geboy	6443 geody	6493 gidiv
6044 fopin	6094 fuday	6144 fuopi	6194 fydof	6244 fyowe	6294 gadiz	6344 ganyl	6394 gebud	6444 geomo	6494 gidro
6045 foplu	6095 fudef	6145 fuovo	6195 fyduc	6245 fypan	6295 gadov	6345 gaoba	6395 gecac	6445 geong	6495 gidud
6046 fopom	6096 fudim	6146 fuoxe	6196 fyeha	6246 fypet	6296 gaemi	6346 gaoci	6396 geceg	6446 geose	6496 gidyt
6047 fopyx	6097 fudoz	6147 fupav	6197 fyelu	6247 fyphy	6297 gaepa	6347 gaoky	6397 gecom	6447 gepad	6497 giebe
6048 forec	6098 fuduv	6148 fupex	6198 fyeny	6248 fypib	6298 gaero	6348 gaord	6398 gecys	6448 gepec	6498 gieju
6049 forol	6099 fuect	6149 fupli	6199 fyepo	6249 fypod	6299 gaest	6349 gaote	6399 gedan	6449 gepij	6499 giend

6500 gierf	6550 gipin	6600 goepu	6650 goops	6700 gucol	6750 gulud	6800 gyatu	6850 gylum	6900 hacym	6950 haogn
6501 gieta	6551 giplu	6601 goery	6651 goorf	6701 gucri	6751 guman	6801 gyawl	6851 gymap	6901 hadet	6951 haoka
6502 giewo	6552 gipom	6602 gofam	6652 gootu	6702 gucub	6752 gumex	6802 gybad	6852 gymes	6902 hadom	6952 haole
6503 gifan	6553 gisaf	6603 gofel	6653 gopaz	6703 gucyn	6753 gumig	6803 gybec	6853 gymiz	6903 hadri	6953 haoro
6504 gifew	6554 gised	6604 gofig	6654 gopep	6704 gudaf	6754 gumoh	6804 gybik	6854 gymof	6904 hadyl	6954 haotz
6505 gifod	6555 gishy	6605 gofot	6655 gopit	6705 guded	6755 gumum	6805 gybol	6855 gymuc	6905 haebi	6955 haoxi
6506 gifri	6556 gisig	6606 gofud	6656 gopok	6706 gudil	6756 gumyl	6806 gybub	6856 gymyt	6906 haeft	6956 hapaj
6507 gifum	6557 gisoc	6607 gogan	6657 gopus	6707 gudoc	6757 gunap	6807 gycaf	6857 gynac	6907 haelk	6957 hapho
6508 gifys	6558 gisuz	6608 gogew	6658 gorab	6708 guduz	6758 gunes	6808 gyced	6858 gyneg	6908 haeme	6958 hapuy
6509 gigiz	6559 gitav	6609 gogod	6659 gorev	6709 gudyx	6759 gunic	6809 gychi	6859 gyniv	6909 haena	6959 haray
6510 gigof	6560 gitem	6610 gogum	6660 goriz	6710 gueba	6760 gunot	6810 gycoc	6860 gynom	6910 haerp	6960 haric
6511 gihak	6561 gitis	6611 gohay	6661 goron	6711 guech	6761 guobu	6811 gycuz	6861 gyoce	6911 haesh	6961 haroh
6512 gihey	6562 gituf	6612 gohet	6662 goruk	6712 guedo	6762 guogy	6812 gycyk	6862 gyodo	6912 hafol	6962 hasib
6513 gihip	6563 giubs	6613 gohor	6663 gosec	6713 guege	6763 guols	6813 gydag	6863 gyoli	6913 hafra	6963 hasko
6514 gihux	6564 giuck	6614 gohun	6664 gosol	6714 gueki	6764 guond	6814 gydeb	6864 gyoms	6914 hafuf	6964 hasla
6515 gijar	6565 giuly	6615 gohym	6665 gospy	6715 guemp	6765 guorc	6815 gydid	6865 gyonz	6915 hagaz	6965 hastu
6516 gijez	6566 giuno	6616 goici	6666 gosub	6716 guent	6766 guoto	6816 gydox	6866 gyoth	6916 hagli	6966 haswe
6517 gijog	6567 giura	6617 goigy	6667 goted	6717 guesu	6767 guoze	6817 gydun	6867 gyowa	6917 hagoc	6967 hasyf
6518 gikas	6568 giuzi	6618 goiku	6668 gotip	6718 gufag	6768 gupar	6818 gyeky	6868 gyoxu	6918 hagre	6968 hatan
6519 gikek	6569 givag	6619 goilo	6669 gotox	6719 gufeb	6769 gupez	6819 gyeld	6869 gyrav	6919 hagud	6969 hatop
6520 gikid	6570 givil	6620 goima	6670 gotuc	6720 gufir	6770 gupik	6820 gyepi	6870 gyrex	6920 hahac	6970 hatux
6521 gikor	6571 givox	6621 goise	6671 gotyt	6721 gufox	6771 gupog	6821 gyeru	6871 gyrip	6921 hahez	6971 haubu
6522 gikug	6572 givun	6622 gojap	6672 goubo	6722 gufun	6772 guras	6822 gyete	6872 gyrul	6922 hahid	6972 haude
6523 gikyp	6573 giwir	6623 gojes	6673 gouka	6723 gugat	6773 gurek	6823 gyfat	6873 gysar	6923 hahob	6973 haufi
6524 gilaw	6574 giwob	6624 gojil	6674 goune	6724 gugef	6774 gurib	6824 gyfeh	6874 gysez	6924 haimy	6974 havon
6525 gileh	6575 gizuk	6625 gojof	6675 govat	6725 gugid	6775 guror	6825 gyfim	6875 gysog	6925 haini	6975 hawaf
6526 gilit	6576 goacy	6626 gojux	6676 govix	6726 gugoz	6776 gurug	6826 gyfoz	6876 gysux	6926 haisk	6976 hawel
6527 gilur	6577 goadu	6627 gokar	6677 govoc	6727 gugul	6777 gusav	6827 gygak	6877 gytaz	6927 haizo	6977 hawip
6528 gilyd	6578 goafa	6628 gokib	6678 gowal	6728 gugyb	6778 gusim	6828 gygem	6878 gytix	6928 hajad	6978 hawog
6529 gimax	6579 goale	6629 gokog	6679 gowiv	6729 guhaw	6779 gusly	6829 gygly	6879 gytus	6929 hajek	6979 hawye
6530 gimer	6580 goani	6630 golas	6680 goxis	6730 guhok	6780 gusop	6830 gygob	6880 gytyr	6930 hajow	6980 haxar
6531 gimib	6581 goaph	6631 golek	6681 goyeb	6731 guhut	6781 gusur	6831 gyguf	6881 gyudi	6931 hajur	6981 haxen
6532 gimov	6582 goarx	6632 golim	6682 goyom	6732 guhyc	6782 guthy	6832 gyhil	6882 gyugu	6932 hakaw	6982 haxim
6533 gimut	6583 gobak	6633 golug	6683 gozag	6733 guibi	6783 gutit	6833 gyhow	6883 gyuko	6933 hakuk	6983 hayas
6534 gimyn	6584 gobem	6634 golyb	6684 gozoz	6734 guicu	6784 gutow	6834 gyibu	6884 gyule	6934 halav	6984 hayef
6535 ginaz	6585 gobid	6635 gomaj	6685 guabo	6735 guiko	6785 gutre	6835 gyige	6885 gyvas	6935 halep	6985 haygo
6536 ginep	6586 gobuf	6636 gomeh	6686 guacs	6736 guing	6786 guver	6836 gyind	6886 gyvib	6936 halir	6986 hayne
6537 ginim	6587 gobyx	6637 gomik	6687 guald	6737 guisy	6787 guvov	6837 gyirk	6887 gyvor	6937 halok	6987 hayum
6538 ginok	6588 gocaw	6638 gomop	6688 guapy	6738 guith	6788 guvup	6838 gyisl	6888 gywah	6938 halyc	6988 hazap
6539 ginus	6589 gocoy	6639 gomur	6689 guarm	6739 gujak	6789 guwaz	6839 gyivo	6889 gywho	6939 hamat	6989 hazif
6540 giofo	6590 gocut	6640 gomyd	6690 guauk	6740 gujem	6790 guwet	6840 gyixa	6890 haabs	6940 hamey	6990 hazus
6541 gioke	6591 godac	6641 gonax	6691 guawn	6741 gujob	6791 guwod	6841 gykam	6891 haajo	6941 hamod	6991 heako
6542 giold	6592 godeg	6642 goner	6692 guaxi	6742 gujuf	6792 guyew	6842 gykel	6892 haake	6942 hamup	6992 heaxa
6543 giont	6593 godin	6643 gonif	6693 gubac	6743 gukal	6793 guyla	6843 gykif	6893 haalu	6943 hanak	6993 heaze
6544 giopu	6594 godul	6644 gonov	6694 gubeg	6744 guken	6794 guzah	6844 gykro	6894 haarc	6944 hanec	6994 hebri
6545 giorb	6595 godyp	6645 gonup	6695 gubif	6745 gukos	6795 guzin	6845 gykud	6895 haavi	6945 haniz	6995 hecho
6546 giovi	6596 goefi	6646 gonyg	6696 gublu	6746 gukyk	6796 gyala	6846 gylan	6896 habex	6946 hanof	6996 hecip
6547 gioxa	6597 goega	6647 gooch	6697 gubro	6747 gulel	6797 gyano	6847 gylet	6897 hacah	6947 hanut	6997 hedah
6548 gipac	6598 goeke	6648 gooja	6698 gucad	6748 gulix	6798 gyare	6848 gylis	6898 hacig	6948 haoby	6998 heeca
6549 gipeg	6599 goelm	6649 goomi	6699 gucec	6749 gulon	6799 gyasi	6849 gylod	6899 hacul	6949 haocs	6999 heegy

heezo hefak hefle hefyr

[redacted]

SECTION A

7000 hegeb	**7050** hiexe	**7100** hoayo	**7150** hoosk	**7200** hugaw	**7250** hybli	**7300** hynah	**7350** ibant	**7400** ibkil	**7450** ibvez
7001 heglo	7051 hifab	7101 hobaw	7151 hoozu	7201 hugro	7251 hybor	7301 hynef	7351 ibase	7401 ibkun	7451 ibvin
7002 hehed	7052 hifev	7102 hoboj	7152 hopef	7202 huhan	7252 hybug	7302 hynib	7352 ibazo	7402 ibkyd	7452 ibvog
7003 hehuw	7053 hifip	7103 hocat	7153 hoply	7203 huhyl	7253 hycav	7303 hynoz	7353 ibbax	7403 iblay	7453 ibwat
7004 heila	7054 hiflu	7104 hocen	7154 hopru	7204 huims	7254 hycit	7304 hynyl	7354 ibber	7404 iblef	7454 ibwid
7005 heipe	7055 hifon	7105 hoclo	7155 horaj	7205 huirp	7255 hycop	7305 hyoba	7355 ibblo	7405 iblig	7455 ibyah
7006 heito	7056 higac	7106 hocuf	7156 horey	7206 huitz	7256 hycur	7306 hyoci	7356 ibbup	7406 ibloz	7456 ibyen
7007 hejop	7057 higeg	7107 hocyr	7157 horix	7207 hujav	7257 hydax	7307 hyoky	7357 ibbyn	7407 ibluv	7457 ibyow
7008 hekoy	7058 higin	7108 hodew	7158 hoser	7208 hujum	7258 hydiz	7308 hyord	7358 ibcaz	7408 ibmak	7458 ibyut
7009 hekra	7059 higom	7109 hodir	7159 hosky	7209 hukay	7259 hydov	7309 hyote	7359 ibcep	7409 ibmem	7459 ibzet
7010 hekuv	**7060** hihad	**7110** hodot	**7160** hospa	**7210** hukex	**7260** hydup	**7310** hyown	**7360** ibcro	**7410** ibmip	**7460** ibzop
7011 helew	7061 hihif	7111 hodum	7161 hosto	7211 huklo	7261 hyemi	7311 hyoxo	7361 ibcus	7411 ibmob	7461 ibzur
7012 henaw	7062 hijeh	7112 hodys	7162 hosuh	7212 hukut	7262 hyest	7312 hypak	7362 ibdab	7412 ibmuf	7462 icaec
7013 henen	7063 hikaf	7113 hoeco	7163 hotar	7213 huley	7263 hyewe	7313 hypem	7363 ibdev	7413 ibmyg	7463 icaha
7014 henuc	7064 hiked	7114 hoers	7164 hotif	7214 hulil	7264 hyfaz	7314 hypid	7364 ibdic	7414 ibnal	7464 icain
7015 heoks	7065 hikli	7115 hofaf	7165 hotom	7215 humew	7265 hyfep	7315 hyplo	7365 ibdon	7415 ibnek	7465 icalp
7016 heona	7066 hikoc	7116 hofli	7166 hotug	7216 hunos	7266 hyfin	7316 hypry	7366 ibdry	7416 ibnix	7466 icamb
7017 hepoh	7067 hiliz	7117 hofre	7167 hotyb	7217 hunyt	7267 hyfly	7317 hypuf	7367 ibduk	7417 ibnyo	7467 icark
7018 hepul	7068 hilun	7118 hoglu	7168 houfy	7218 huozo	7268 hyfok	7318 hyral	7368 ibebu	7418 ibobs	7468 icayo
7019 herer	7069 hilyt	7119 hoham	7169 houla	7219 hupla	7269 hyfus	7319 hyren	7369 ibefa	7419 iboda	7469 icbaw
7020 herha	**7070** himef	**7120** hoheb	**7170** houpi	**7220** hupof	**7270** hygab	**7320** hyrij	**7370** ibele	**7420** ibofi	**7470** icboj
7021 heryp	7071 himic	7121 hohog	7171 houvu	7221 hurah	7271 hygev	7321 hyros	7371 ibens	7421 ibojo	7471 iccat
7022 hesey	7072 himoz	7122 hohud	7172 hovep	7222 hurog	7272 hygis	7322 hyrut	7372 ibeph	7422 iboku	7472 iccen
7023 hesma	7073 hinem	7123 hoibo	7173 hovod	7223 hurys	7273 hygon	7323 hysam	7373 iberd	7423 ibowl	7473 icclo
7024 hetib	7074 hinid	7124 hoile	7174 howeg	7224 husak	7274 hyhap	7324 hysel	7374 ibewi	7424 ibpam	7474 iccuf
7025 hetok	7075 hiofu	7125 hoink	7175 howov	7225 hushu	7275 hyhew	7325 hysho	7375 ibfac	7425 ibpel	7475 iccyr
7026 hetud	7076 hioly	7126 hoixu	7176 hoxan	7226 husme	7276 hyhof	7326 hysif	7376 ibfeg	7426 ibpiv	7476 icdew
7027 hetyg	7077 hiomp	7127 hojip	7177 hoxor	7227 hutek	7277 hyhuc	7327 hysud	7377 ibfib	7427 ibpot	7477 icdir
7028 hewif	7078 hiort	7128 hojok	7178 hoyac	7228 hutho	7278 hyibs	7328 hysyk	7378 ibfom	7428 ibpud	7478 icdum
7029 hexav	7079 hiosi	7129 hojun	7179 hoyol	7229 huvur	7279 hyice	7329 hytat	7379 ibfru	7429 ibran	7479 icdys
7030 heyun	**7080** hiowo	**7130** hokah	**7180** hozak	**7230** huwam	**7280** hyiod	**7330** hytez	**7380** ibgad	**7430** ibrew	**7480** iceco
7031 hezar	7081 hirez	7131 hoket	7181 hozem	7231 huwer	7281 hyipi	7331 hytig	7381 ibgec	7431 ibrik	7481 iceig
7032 hezeg	7082 hiryn	7132 hokow	7182 hoziz	7232 huwiz	7282 hyirl	7332 hytoy	7382 ibgif	7432 ibrod	7482 icers
7033 hezis	7083 hitas	7133 hokri	7183 hozob	7233 huxyn	7283 hykad	7333 hytru	7383 ibgol	7433 ibrum	7483 iceux
7034 hiabi	7084 hitey	7134 hokyx	7184 hozup	7234 huyel	7284 hykec	7334 hyufo	7384 ibgub	7434 ibryl	7484 icfaf
7035 hiady	7085 hiufa	7135 holaz	7185 huary	7235 huzas	7285 hykir	7335 hyuma	7385 ibhav	7435 ibsap	7485 icfli
7036 hiama	7086 hiunc	7136 holic	7186 hubeb	7236 huzez	7286 hykol	7336 hyund	7386 ibhit	7436 ibses	7486 icfos
7037 hiapo	7087 hiurg	7137 holyl	7187 hubir	7237 huzor	7287 hykub	7337 hyuse	7387 ibhoy	7437 ibsiz	7487 icfre
7038 hibav	7088 hiust	7138 homay	7188 hubul	7238 huzuc	7288 hylaf	7338 hyuti	7388 ibhul	7438 ibsof	7488 icglu
7039 hibij	7089 hiuxo	7139 homis	7189 hucef	7239 hyabu	7289 hyled	7339 hyvan	7389 ibibe	7439 ibsuc	7489 icham
7040 hiboh	**7090** hiuze	**7140** homul	**7190** hucim	**7240** hyaca	**7290** hylip	**7340** hyvik	**7390** ibica	**7440** ibsyr	**7490** icheb
7041 hicib	7091 hiviv	7141 honex	7191 hudec	7241 hyads	7291 hyloc	7341 hyvum	7391 ibifo	7441 ibtaw	7491 ichog
7042 hicov	7092 hiwor	7142 honil	7192 hudym	7242 hyafi	7292 hyluz	7342 hywar	7392 ibigu	7442 ibtex	7492 ichud
7043 hidaz	7093 hixel	7143 honoy	7193 hueda	7243 hyago	7293 hylyn	7343 hywes	7393 ibimp	7443 ibthi	7493 iciap
7044 hidep	7094 hizek	7144 honub	7194 huemb	7244 hyalm	7294 hymag	7344 hywot	7394 ibits	7444 ibtuy	7494 icibo
7045 hidok	7095 hizir	7145 hoofa	7195 hueni	7245 hyane	7295 hymeb	7345 ibaba	7395 ibjaf	7445 ibuga	7495 icink
7046 hiedi	7096 hoaha	7146 hooge	7196 hueps	7246 hyaty	7296 hymil	7346 ibach	7396 ibjed	7446 ibumi	7496 icjip
7047 hierl	7097 hoalp	7147 hoolm	7197 huetu	7247 hyawk	7297 hymox	7347 ibaim	7397 ibjoc	7447 iburn	7497 icjok
7048 hiesy	7098 hoamb	7148 hoopo	7198 huexo	7248 hybas	7298 hymun	7348 ibaju	7398 ibkag	7448 ibutu	7498 icjun
7049 hiewa	7099 hoark	7149 hoorn	7199 hufis	7249 hybek	7299 hymyc	7349 ibaly	7399 ibkeb	7449 ibvar	7499 ickah

7500 icket	**7550** icxan	**7600** idisa	**7650** idtum	**7700** ifhuf	**7750** iftah	**7800** igewe	**7850** igpem	**7900** iheky	**7950** ihoxu
7501 ickow	7551 icxor	7601 idivy	7651 idtyx	7701 ifhyg	7751 iftim	7801 igfaz	7851 igpid	7901 ihepi	7951 ihpax
7502 ickri	7552 icyol	7602 idjan	7652 iducu	7702 ifiba	7752 iftob	7802 igfep	7852 igplo	7902 iheru	7952 ihper
7503 ickyx	7553 iczak	7603 idjeb	7653 iduki	7703 ifife	7753 iftyn	7803 igfin	7853 igpry	7903 ihfat	7953 ihpic
7504 iclaz	7554 iczem	7604 idjik	7654 idung	7704 ifimi	7754 ifuco	7804 igfly	7854 igpuf	7904 ihfeh	7954 ihpov
7505 iclic	7555 icziz	7605 idkap	7655 idupa	7705 ifiru	7755 ifuda	7805 igfok	7855 igral	7905 ihfim	7955 ihpup
7506 iclus	7556 iczup	7606 idkes	7656 iduso	7706 ifiso	7756 ifuky	7806 igfus	7856 igren	7906 ihfoz	7956 ihrav
7507 iclyl	7557 idadz	7607 idkim	7657 idvad	7707 ifjas	7757 ifume	7807 iggab	7857 igrij	7907 ihgak	7957 ihrex
7508 icmay	7558 idalc	7608 idkuc	7658 idvec	7708 ifjet	7758 ifuph	7808 iggev	7858 igros	7908 ihgem	7958 ihrip
7509 icmis	7559 idana	7609 idlar	7659 idvir	7709 ifjor	7759 ifuvi	7809 iggis	7859 igrut	7909 ihgly	7959 ihrot
7510 icmul	**7560** idari	**7610** idlez	**7660** idvol	**7710** ifkav	**7760** ifuzu	**7810** iggon	**7860** igsam	**7910** ihgob	**7960** ihrym
7511 icnex	7561 idasm	7611 idlid	7661 idwac	7711 ifkew	7761 ifvay	7811 ighap	7861 igsel	7911 ihguf	7961 ihsar
7512 icnil	7562 idaxu	7612 idlog	7662 idwed	7712 ifkiz	7762 ifvef	7812 ighew	7862 igsho	7912 ihibu	7962 ihsin
7513 icnoy	7563 idbag	7613 idlux	7663 idyls	7713 ifkry	7763 ifvid	7813 ighof	7863 igsif	7913 ihify	7963 ihsux
7514 icnub	7564 idbef	7614 idlym	7664 idyoc	7714 iflax	7764 ifvoz	7814 ighuc	7864 igsud	7914 ihige	7964 ihtaz
7515 icofa	7565 idbib	7615 idmas	7665 idzom	7715 ifler	7765 ifwem	7815 ighyd	7865 igsyk	7915 ihind	7965 ihtix
7516 icoge	7566 idbox	7616 idmek	7666 ifaci	7716 iflup	7766 ifwis	7816 igibs	7866 igtat	7916 ihirk	7966 ihtyr
7517 icolm	7567 idbun	7617 idmit	7667 ifado	7717 ifmaz	7767 ifyak	7817 igice	7867 igtez	7917 ihivo	7967 ihudi
7518 icopo	7568 idbyl	7618 idmor	7668 ifafu	7718 ifmep	7768 ifzal	7818 igiod	7868 igtig	7918 ihixa	7968 ihugu
7519 icorn	7569 idcay	7619 idmug	7669 ifaga	7719 ifmin	7769 ifzen	7819 igipi	7869 igtoy	7919 ihjal	7969 ihuko
7520 icozu	**7570** idcet	**7620** idnav	**7670** ifalk	**7720** ifmok	**7770** ifzip	**7820** igirl	**7870** igtru	**7920** ihjen	**7970** ihuna
7521 icply	7571 idcoz	7621 idneh	7671 ifbam	7721 ifmus	7771 ifzug	7821 igkad	7871 igufo	7921 ihjos	7971 ihvas
7522 icpru	7572 iddak	7622 idnop	7672 ifbel	7722 ifnab	7772 igabu	7822 igkec	7872 iguma	7922 ihjut	7972 ihvej
7523 icraj	7573 iddem	7623 idnur	7673 ifbiv	7723 ifnev	7773 igaca	7823 igkir	7873 igund	7923 ihkam	7973 ihvib
7524 icrey	7574 iddig	7624 idobe	7674 ifboy	7724 ifnig	7774 igads	7824 igkol	7874 iguse	7924 ihkel	7974 ihvor
7525 icrix	7575 iddob	7625 idock	7675 ifbud	7725 ifnon	7775 igafi	7825 igkub	7875 iguti	7925 ihkif	7975 ihwah
7526 icser	7576 idduf	7626 idolo	7676 ifcac	7726 ifnuk	7776 igago	7826 iglaf	7876 igvan	7926 ihkro	7976 ihwey
7527 icsid	7577 iddyn	7627 idomu	7677 ifceg	7727 ifnyp	7777 igalm	7827 igled	7877 igvik	7927 ihkud	7977 ihwho
7528 icsky	7578 idebo	7628 idoni	7678 ifcom	7728 ifody	7778 igane	7828 iglip	7878 igvum	7928 ihlan	7978 ihyug
7529 icspa	7579 idede	7629 idopy	7679 ifcys	7729 ifola	7779 igaty	7829 igloc	7879 igwar	7929 ihlet	7979 ijabo
7530 icsto	**7580** idefy	**7630** idota	**7680** ifdan	**7730** ifomo	**7780** igawk	**7830** igluz	**7880** igwes	**7930** ihlis	**7980** ijacs
7531 icsuh	7581 ideku	7631 idpab	7681 ifdey	7731 ifong	7781 igbas	7831 iglyn	7881 ihano	7931 ihlod	7981 ijapy
7532 ictar	7582 idern	7632 idpev	7682 ifdif	7732 ifose	7782 igbek	7832 igmag	7882 ihare	7932 ihlum	7982 ijarm
7533 ictel	7583 ideti	7633 idpon	7683 ifdod	7733 ifpad	7783 igbli	7833 igmeb	7883 ihasi	7933 ihmap	7983 ijate
7534 ictif	7584 idfaw	7634 idpuk	7684 ifdut	7734 ifpec	7784 igbor	7834 igmil	7884 ihatu	7934 ihmes	7984 ijauk
7535 ictom	7585 idfen	7635 idpyc	7685 ifecu	7735 ifpij	7785 igbug	7835 igmox	7885 ihawl	7935 ihmiz	7985 ijawn
7536 ictug	7586 idfic	7636 idrax	7686 ifego	7736 ifpol	7786 igcav	7836 igmun	7886 ihbad	7936 ihmof	7986 ijaxi
7537 ictyb	7587 idfoy	7637 idrhe	7687 ifeja	7737 ifpub	7787 igcit	7837 igmyc	7887 ihbec	7937 ihmuc	7987 ijaza
7538 icufy	7588 idfut	7638 idrov	7688 ifeli	7738 ifpyr	7788 igcop	7838 ignef	7888 ihbik	7938 ihmyt	7988 ijbac
7539 icula	7589 idgam	7639 idrup	7689 iferx	7739 ifraf	7789 igcur	7839 ignib	7889 ihbol	7939 ihnac	7989 ijbeg
7540 icumo	**7590** idgel	**7640** idryd	**7690** ifeve	**7740** ifred	**7790** igcyg	**7840** ignoz	**7890** ihbub	**7940** ihneg	**7990** ijbif
7541 icupi	7591 idgip	7641 idsaz	7691 ifews	7741 ifril	7791 igdax	7841 ignyl	7891 ihcaf	7941 ihniv	7991 ijblu
7542 icuvu	7592 idgot	7642 idsep	7692 iffap	7742 ifroc	7792 igder	7842 igoba	7892 ihced	7942 ihnom	7992 ijbro
7543 icvep	7593 idgru	7643 idsmi	7693 iffit	7743 ifruz	7793 igdiz	7843 igoci	7893 ihcuz	7943 ihnys	7993 ijcad
7544 icvod	7594 idhaf	7644 idsty	7694 iffro	7744 ifsag	7794 igdov	7844 igoky	7894 ihcyk	7944 ihoce	7994 ijcec
7545 icwad	7595 idher	7645 idsus	7695 ifgar	7745 ifseb	7795 igdup	7845 igord	7895 ihdag	7945 ihodo	7995 ijcol
7546 icweg	7596 idhis	7646 idtal	7696 ifgez	7746 ifsik	7796 igemi	7846 igote	7896 ihdeb	7946 iholi	7996 ijcri
7547 icwhy	7597 idhuz	7647 idteg	7697 ifgog	7747 ifsox	7797 igepa	7847 igown	7897 ihdid	7947 ihoms	7997 ijcub
7548 icwim	7598 idifu	7648 idtin	7698 ifhat	7748 ifsun	7798 igero	7848 igoxo	7898 ihdox	7948 ihonz	7998 ijcyn
7549 icwov	7599 idilk	7649 idtos	7699 ifhib	7749 ifsyd	7799 igest	7849 igpak	7899 ihdun	7949 ihowa	7999 ijdaf

ijded ijdil ijdoc ijduz

[redacted]

SECTION A

8000 ijdyx	**8050** ijrib	**8100** ikili	**8150** ilawe	**8200** illeg	**8250** imbuz	**8300** imsev	**8350** inmow	**8400** ipeke	**8450** ipoch
8001 ijeba	8051 ijror	8101 ikimu	8151 ilaya	8201 illif	8251 imceb	8301 imsis	8351 inmuy	8401 ipelm	8451 ipode
8002 ijech	8052 ijrug	8102 ikina	8152 ilbap	8202 illom	8252 imcil	8302 imsym	8352 innaw	8402 ipepu	8452 ipoja
8003 ijedo	8053 ijsav	8103 ikipo	8153 ilbes	8203 illyp	8253 imcyd	8303 imtyp	8353 innuc	8403 ipery	8453 ipomi
8004 ijege	8054 ijsim	8104 ikist	8154 ilbig	8204 ilmad	8254 imdef	8304 imubi	8354 inoks	8404 ipfam	8454 ipops
8005 ijeki	8055 ijsly	8105 ikjab	8155 ilbuc	8205 ilmec	8255 imdoz	8305 imucy	8355 inona	8405 ipfel	8455 iporf
8006 ijemp	8056 ijsop	8106 ikjif	8156 ilbyb	8206 ilmix	8256 imduv	8306 imuge	8356 inpat	8406 ipfig	8456 ipotu
8007 ijent	8057 ijsur	8107 ikjon	8157 ilcar	8207 ilmol	8257 imect	8307 imulo	8357 inpoh	8407 ipfot	8457 ippaz
8008 ijesu	8058 ijtab	8108 ikkac	8158 ilchy	8208 ilnaf	8258 imegu	8308 imups	8358 inrer	8408 ipfud	8458 ippep
8009 ijfag	8059 ijthy	8109 ikkyn	8159 ilcic	8209 ilned	8259 imeko	8309 imuxu	8359 inrha	8409 ipgan	8459 ippit
8010 ijfeb	**8060** ijtit	**8110** iklec	**8160** ilcog	**8210** ilnoc	**8260** imems	**8310** imuza	**8360** inrim	**8410** ipgew	**8460** ippok
8011 ijfir	8061 ijtow	8111 ikliv	8161 ilcux	8211 ilnuz	8261 imepy	8311 imvac	8361 inryp	8411 ipgic	8461 ippus
8012 ijfox	8062 ijube	8112 iklol	8162 ildas	8212 ilodi	8262 imexa	8312 imveg	8362 insey	8412 ipgod	8462 iprab
8013 ijfun	8063 ijuja	8113 iklub	8163 ildek	8213 ilonu	8263 imezi	8313 imvom	8363 inshi	8413 ipgum	8463 iprev
8014 ijgat	8064 ijult	8114 ikmaf	8164 ildor	8214 ilost	8264 imfah	8314 imwak	8364 insma	8414 iphay	8464 ipriz
8015 ijgef	8065 ijumy	8115 ikmed	8165 ildug	8215 iloxy	8265 imfem	8315 imwic	8365 insot	8415 iphor	8465 ipron
8016 ijgid	8066 ijunz	8116 ikmoc	8166 ildyc	8216 ilpag	8266 imgen	8316 imzer	8366 intib	8416 iphun	8466 ipruk
8017 ijgoz	8067 ijvov	8117 ikmuz	8167 ileid	8217 ilpeb	8267 imgiv	8317 inako	8367 intok	8417 iphym	8467 ipsad
8018 ijgul	8068 ijvup	8118 iknox	8168 ilelz	8218 ilpim	8268 imgos	8318 inats	8368 intyg	8418 ipici	8468 ipsec
8019 ijhok	8069 ijwaz	8119 ikodu	8169 ilept	8219 ilpow	8269 imheh	8319 inaug	8369 inwap	8419 ipigy	8469 ipspy
8020 ijhyc	**8070** ijwet	**8120** ikogo	**8170** ilere	**8220** ilpun	**8270** imhuy	**8320** inaxa	**8370** inwif	**8420** ipiku	**8470** ipsti
8021 ijkal	8071 ijwod	8121 ikoha	8171 ilevu	8221 ilraw	8271 imily	8321 inaze	8371 inxav	8421 ipilo	8471 ipsub
8022 ijkos	8072 ikanz	8122 ikove	8172 ilfav	8222 ilref	8272 imino	8322 inbay	8372 inyab	8422 ipima	8472 ipted
8023 ijkyk	8073 ikapa	8123 ikoys	8173 ilfey	8223 ilroz	8273 imize	8323 inbom	8373 inyes	8423 ipise	8473 iptip
8024 ijlam	8074 ikask	8124 ikpir	8174 ilfiz	8224 ilsah	8274 imjel	8324 inbri	8374 inzar	8424 ipjap	8474 iptox
8025 ijlel	8075 ikath	8125 ikpoz	8175 ilfop	8225 ilsem	8275 imjig	8325 incip	8375 inzeg	8425 ipjes	8475 iptuc
8026 ijlix	8076 ikauc	8126 ikrak	8176 ilfur	8226 ilsip	8276 imkod	8326 incly	8376 inzis	8426 ipjil	8476 iptyt
8027 ijlon	8077 ikaxo	8127 ikrem	8177 ilgax	8227 ilsob	8277 imkum	8327 incru	8377 ipacy	8427 ipjof	8477 ipubo
8028 ijlud	8078 ikbez	8128 ikrob	8178 ilger	8228 ilsuf	8278 imlik	8328 indah	8378 ipadu	8428 ipjux	8478 ipuka
8029 ijmex	8079 ikbim	8129 ikruf	8179 ilgib	8229 iltam	8279 imluc	8329 indit	8379 ipafa	8429 ipkar	8479 ipulu
8030 ijmig	**8080** ikbly	**8130** ikscu	**8180** ilgov	**8230** iltep	**8280** immog	**8330** indye	**8380** ipale	**8430** ipkib	**8480** ipune
8031 ijmoh	8081 ikcin	8131 ikski	8181 ilgup	8231 iltiv	8281 immux	8331 ineca	8381 ipani	8431 ipkog	8481 ipvat
8032 ijmum	8082 ikcor	8132 iksos	8182 ilhaj	8232 iltot	8282 imnas	8332 inegy	8382 ipaph	8432 iplas	8482 ipvix
8033 ijmyl	8083 ikcug	8133 iktiz	8183 ilhir	8233 iltul	8283 imnor	8333 ineum	8383 iparx	8433 iplek	8483 ipvoc
8034 ijnap	8084 ikcyb	8134 iktur	8184 ilhox	8234 iludy	8284 imnug	8334 inezo	8384 ipato	8434 iplim	8484 ipwal
8035 ijnic	8085 ikdav	8135 iktyl	8185 ilhuk	8235 ilval	8285 imnyx	8335 infak	8385 ipbak	8435 iplow	8485 ipwiv
8036 ijnot	8086 ikdej	8136 ikuca	8186 ilhys	8236 ilven	8286 imoga	8336 infle	8386 ipbem	8436 iplug	8486 ipxis
8037 ijobu	8087 ikdop	8137 ikunt	8187 ilics	8237 ilvit	8287 imolu	8337 infyr	8387 ipbid	8437 iplyb	8487 ipyeb
8038 ijogy	8088 ikebs	8138 ikvam	8188 iligo	8238 ilvos	8288 imopi	8338 ingeb	8388 ipbob	8438 ipmaj	8488 ipyom
8039 ijols	8089 ikegi	8139 ikvel	8189 ilika	8239 ilway	8289 imovo	8339 inglo	8389 ipbuf	8439 ipmeh	8489 ipzag
8040 ijond	**8090** ikema	**8140** ikvij	**8190** iliph	**8240** ilwex	**8290** imoxe	**8340** inhas	**8390** ipbyx	**8440** ipmik	**8490** ipzoz
8041 ijorc	8091 ikeno	8141 ikvot	8191 ility	8241 ilxin	8291 impav	8341 inila	8391 ipcaw	8441 ipmop	8491 iqabi
8042 ijoto	8092 ikepe	8142 ikyat	8192 ilizu	8242 ilzan	8292 impex	8342 inipe	8392 ipcoy	8442 ipmur	8492 iqact
8043 ijoze	8093 ikert	8143 ikzeh	8193 iljat	8243 ilzik	8293 impyk	8343 injam	8393 ipcut	8443 ipmyd	8493 iqage
8044 ijpar	8094 ikezu	8144 ikzof	8194 iljew	8244 ilzum	8294 imraz	8344 injop	8394 ipdac	8444 ipnax	8494 iqama
8045 ijpez	8095 ikgep	8145 ilahs	8195 ilkev	8245 imang	8295 imrif	8345 inkoy	8395 ipdeg	8445 ipner	8495 iqapo
8046 ijpik	8096 ikgil	8146 ilaki	8196 ilkis	8246 imaul	8296 imrok	8346 inkra	8396 ipdin	8446 ipnif	8496 iqaux
8047 ijpog	8097 ikgus	8147 ilalb	8197 ilklu	8247 imbaf	8297 imrud	8347 inkuv	8397 ipdul	8447 ipnov	8497 iqbav
8048 ijras	8098 ikhaz	8148 ilamt	8198 ilkon	8248 imboc	8298 imryc	8348 inlew	8398 ipdyp	8448 ipnup	8498 iqble
8049 ijrek	8099 ikhyp	8149 ilaud	8199 ilkyr	8249 imbry	8299 imsab	8349 inmet	8399 ipefi	8449 ipnyg	8499 iqboh

8500 iqbur	**8550** iqosi	**8600** irexo	**8650** irxem	**8700** islok	**8750** iszif	**8800** itkep	**8850** itzoy	**8900** ivjog	**8950** ivubs
8501 iqcax	8551 iqowo	8601 irfed	8651 irxyn	8701 islyc	8751 itabe	8801 itkin	8851 ivaco	8901 ivkas	8951 ivuck
8502 iqcer	8552 iqpal	8602 irfis	8652 iryap	8702 ismat	8752 itack	8802 itkok	8852 ivada	8902 ivkek	8952 ivuly
8503 iqcib	8553 iqpen	8603 irgaw	8653 iryel	8703 ismey	8753 itaft	8803 itkus	8853 ivaik	8903 ivkid	8953 ivuno
8504 iqcov	8554 iqpig	8604 irget	8654 irzor	8704 ismod	8754 itagi	8804 itlal	8854 ivans	8904 ivkor	8954 ivura
8505 iqcup	8555 iqpos	8605 irgig	8655 irzuc	8705 ismup	8755 itahu	8805 itlex	8855 ivarg	8905 ivkug	8955 ivuzi
8506 iqdaz	8556 iqrat	8606 irgro	8656 isaam	8706 isnak	8756 itaja	8806 itlib	8856 ivavu	8906 ivkyp	8956 ivvag
8507 iqdep	8557 iqrow	8607 irguy	8657 isabs	8707 isnec	8757 itamy	8807 itloh	8857 ivbal	8907 ivlaw	8957 ivvil
8508 iqdix	8558 iqryn	8608 irhan	8658 isajo	8708 isniz	8758 itarn	8808 itlut	8858 ivben	8908 ivleh	8958 ivvox
8509 iqdok	8559 iqsew	8609 irhoc	8659 isake	8709 isnof	8759 itasp	8809 itlyr	8859 ivbix	8909 ivlit	8959 ivvun
8510 iqdus	**8560** iqsil	**8610** irhyl	**8660** isalu	**8710** isnut	**8760** itawo	**8810** itmac	**8860** ivbos	**8910** ivlur	**8960** ivwir
8511 iqedi	8561 iqsod	8611 irims	8661 isarc	8711 isoby	8761 itban	8811 itmeg	8861 ivcab	8911 ivlyd	8961 ivwob
8512 iqemo	8562 iqsum	8612 irirp	8662 isavi	8712 isocs	8762 itbet	8812 itmom	8862 ivcli	8912 ivmax	8962 ivyet
8513 iqerl	8563 iqtas	8613 irjod	8663 isbex	8713 isogn	8763 itbiz	8813 itnik	8863 ivcon	8913 ivmer	8963 ivzic
8514 iqesy	8564 iqtey	8614 irjum	8664 iscah	8714 isoka	8764 itbum	8814 itnol	8864 ivcre	8914 ivmib	8964 ivzuk
8515 iqfab	8565 iqtik	8615 irkay	8665 iscig	8715 isoro	8765 itcap	8815 itobo	8865 ivdam	8915 ivmov	8965 ixaka
8516 iqfev	8566 iqtut	8616 irkex	8666 iscul	8716 isotz	8766 itces	8816 itofy	8866 ivdel	8916 ivmut	8966 ixalo
8517 iqfip	8567 iqufa	8617 irklo	8667 iscym	8717 ispaj	8767 itciv	8817 itome	8867 ivdiv	8917 ivmyn	8967 ixamp
8518 iqflu	8568 iqunc	8618 irkut	8668 isdet	8718 ispho	8768 itcuc	8818 itonc	8868 ivdro	8918 ivnaz	8968 ixanu
8519 iqfon	8569 iqurg	8619 irlab	8669 isdom	8719 ispuy	8769 itcyx	8819 itopa	8869 ivdud	8919 ivnep	8969 ixard
8520 iqhad	**8570** iqust	**8620** irley	**8670** isdyl	**8720** isray	**8770** itdar	**8820** itori	**8870** ivdyt	**8920** ivnim	**8970** ixasy
8521 iqhif	8571 iquxo	8621 irlil	8671 isebi	8721 isric	8771 itdez	8821 itots	8871 ivebe	8921 ivnok	8971 ixaxt
8522 iqhot	8572 iquze	8622 irlop	8672 iseft	8722 isroh	8772 itdip	8822 itpaf	8872 ivecs	8922 ivnus	8972 ixbah
8523 iqhug	8573 iqvap	8623 irluk	8673 iselk	8723 issib	8773 itdog	8823 itped	8873 iveju	8923 ivofo	8973 ixbey
8524 iqigi	8574 iqves	8624 irmew	8674 iseme	8724 issko	8774 itdux	8824 itpil	8874 ivend	8924 ivoke	8974 ixbot
8525 iqinz	8575 iqviv	8625 irmoy	8675 isena	8725 issla	8775 itedy	8825 itpoc	8875 iverf	8925 ivold	8975 ixcaj
8526 iqism	8576 iqwit	8626 irnin	8676 iserp	8726 isstu	8776 itefo	8826 itpuz	8876 iveta	8926 ivont	8976 ixcid
8527 iqkaf	8577 iqwor	8627 irnos	8677 isesh	8727 issyf	8777 itelv	8827 itrag	8877 ivewo	8927 ivopu	8977 ixcos
8528 iqked	8578 irand	8628 irnyt	8678 iseuc	8728 istan	8778 itemu	8828 itreb	8878 ivfan	8928 ivorb	8978 ixdat
8529 iqkli	8579 iraus	8629 iroaf	8679 isfol	8729 istop	8779 iterg	8829 itrid	8879 ivfew	8929 ivovi	8979 ixdru
8530 iqkoc	**8580** irbat	**8630** iroen	**8680** isfra	**8730** istux	**8780** itese	**8830** itrox	**8880** ivfri	**8930** ivoxa	**8980** ixean
8531 iqlag	8581 irbeb	8631 iroid	8681 isfuf	8731 isual	8781 itfas	8831 itrun	8881 ivfys	8931 ivpac	8981 ixefu
8532 iqleb	8582 irbir	8632 irora	8682 isgli	8732 isubu	8782 itfug	8832 itryf	8882 ivgap	8932 ivpeg	8982 ixely
8533 iqliz	8583 irbow	8633 iroux	8683 isgoc	8733 isude	8783 itgav	8833 itsit	8883 ivges	8933 ivpin	8983 ixerk
8534 iqlox	8584 irbul	8634 irows	8684 isgre	8734 isufi	8784 itgim	8834 itsys	8884 ivgiz	8934 ivplu	8984 ixevo
8535 iqlun	8585 ircal	8635 irozo	8685 isgud	8735 isuor	8785 itgop	8835 ittha	8885 ivgof	8935 ivpom	8985 ixfal
8536 iqlyt	8586 ircef	8636 irpip	8686 ishez	8736 isvis	8786 itgur	8836 ittry	8886 ivguc	8936 ivpyx	8986 ixfet
8537 iqmah	8587 ircim	8637 irpla	8687 isimy	8737 isvon	8787 itgyn	8837 ittuk	8887 ivgym	8937 ivrad	8987 ixfif
8538 iqmef	8588 ircot	8638 irpof	8688 isini	8738 iswaf	8788 ithab	8838 ituba	8888 ivhak	8938 ivrec	8988 ixfoc
8539 iqmic	8589 irdad	8639 irpre	8689 isjad	8739 iswip	8789 ithek	8839 itufe	8889 ivhey	8939 ivrol	8989 ixgeh
8540 iqmoz	**8590** irdec	**8640** irpuh	**8690** isjek	**8740** iswog	**8790** ithig	**8840** ituku	**8890** ivhip	**8940** ivrub	**8990** ixgow
8541 iqnem	8591 irdol	8641 irsak	8691 isjow	8741 iswye	8791 ithon	8841 ituni	8891 ivhoz	8941 ivsaf	8991 ixgra
8542 iqnid	8592 irdub	8642 irshu	8692 isjur	8742 isyas	8792 ithup	8842 ituzo	8892 ivhux	8942 ivsed	8992 ixhag
8543 iqnob	8593 irdym	8643 irtag	8693 iskaw	8743 isxar	8793 itiny	8843 itvak	8893 ivich	8943 ivshy	8993 ixhem
8544 iqnuf	8594 ireda	8644 irtek	8694 iskem	8744 isxen	8794 itipu	8844 itvem	8894 ivido	8944 ivsig	8994 ixhiz
8545 iqoak	8595 iremb	8645 irtho	8695 iskot	8745 isxim	8795 ititi	8845 itvic	8895 ivine	8945 ivsoc	8995 ixhop
8546 iqofu	8596 ireni	8646 irtuz	8696 iskuk	8746 isyef	8796 itiwa	8846 itweh	8896 ivisi	8946 ivsuz	8996 ixhus
8547 iqomp	8597 ireps	8647 irupu	8697 islav	8747 isygo	8797 itjer	8847 ityub	8897 iviza	8947 ivtav	8997 ixhyb
8548 iqone	8598 irerz	8648 irvur	8698 islep	8748 isyne	8798 itjis	8848 itzam	8898 ivjar	8948 ivtem	8998 ixiar
8549 iqort	8599 iretu	8649 irwer	8699 islir	8749 isyum	8799 itkaz	8849 itzel	8899 ivjez	8949 ivtuf	8999 ixiky

ixilu ixins ixipa ixixi

[redacted]

SECTION A

9000 ixjaz	**9050** ixuve	**9100** izhiv	**9150** izujo	**9200** jahec	**9250** jaujo	**9300** jefub	**9350** jesek	**9400** jifes	**9450** jivef
9001 ixjec	9051 ixuxa	9101 izids	9151 izulk	9201 jahiv	9251 jaulk	9301 jegag	9351 jesor	9401 jifit	9451 jivid
9002 ixjin	9052 ixvex	9102 iziga	9152 izvaz	9202 jaids	9252 javaz	9302 jegir	9352 jesug	9402 jifro	9452 jivoz
9003 ixjol	9053 ixvig	9103 izime	9153 izvey	9203 jaiga	9253 javey	9303 jegle	9353 jesyl	9403 jiful	9453 jiwaw
9004 ixjud	9054 ixvyn	9104 izinu	9154 izvim	9204 jaime	9254 javim	9304 jegun	9354 jetac	9404 jigar	9454 jiwem
9005 ixkik	9055 ixwab	9105 iziri	9155 izvok	9205 jainu	9255 javok	9305 jehah	9355 jetil	9405 jigez	9455 jiwis
9006 ixkof	9056 ixwen	9106 izixo	9156 izvuf	9206 jairi	9256 javuf	9306 jeheg	9356 jetof	9406 jigog	9456 jizal
9007 ixkux	9057 ixwil	9107 izkef	9157 izwag	9207 jaixo	9257 jawag	9307 jehur	9357 jeugo	9407 jihat	9457 jizen
9008 ixlak	9058 ixwox	9108 izkoz	9158 izweb	9208 jajaw	9258 jaweb	9308 jeiby	9358 jeuli	9408 jihib	9458 jizip
9009 ixlev	9059 ixyez	9109 izkru	9159 izwin	9209 jakef	9259 jawin	9309 jeico	9359 jeumu	9409 jihuf	9459 jizug
9010 ixlor	**9060** ixyok	**9110** izlah	**9160** izwow	**9210** jakoz	**9260** jawow	**9310** jeifa	**9360** jeush	**9410** jihyg	**9460** joafo
9011 ixlys	9061 ixyuc	9111 izlen	9161 izwry	9211 jakru	9261 jawry	9311 jeils	9361 jeuta	9411 jikav	9461 joahy
9012 ixmaw	9062 ixzac	9112 izlos	9162 izxal	9212 jalah	9262 jaxal	9312 jeitu	9362 jevav	9412 jikew	9462 joang
9013 ixmyx	9063 ixzep	9113 izluw	9163 izzig	9213 jalen	9263 jaxur	9313 jejef	9363 jevet	9413 jikiz	9463 joape
9014 ixnad	9064 ixzim	9114 izmam	9164 izzol	9214 jalos	9264 jayaf	9314 jejid	9364 jevif	9414 jikry	9464 joara
9015 ixnir	9065 ixzon	9115 izmel	9165 jaacu	9215 jaluw	9265 jayed	9315 jejot	9365 jevoy	9415 jilax	9465 jobaf
9016 ixnul	9066 ixzuf	9116 izmir	9166 jaams	9216 jamam	9266 jayop	9316 jejup	9366 jewom	9416 jiler	9466 jobed
9017 ixony	9067 izacu	9117 izmot	9167 jaaro	9217 jamel	9267 jazad	9317 jekak	9367 jewyn	9417 jilot	9467 jobit
9018 ixoph	9068 izams	9118 izmud	9168 jaata	9218 jamir	9268 jazig	9318 jekeh	9368 jeyem	9418 jilup	9468 joboc
9019 ixore	9069 izaro	9119 iznar	9169 jaave	9219 jamot	9269 jazol	9319 jekob	9369 jeyux	9419 jimaz	9469 jobry
9020 ixoso	**9070** izave	**9120** iznez	**9170** jaaxy	**9220** jamud	**9270** jazub	**9320** jekuf	**9370** jezaf	**9420** jimep	**9470** jobuz
9021 ixout	9071 izaxy	9121 iznip	9171 jaazi	9221 janar	9271 jeade	9321 jekys	9371 jezed	9421 jimok	9471 jocag
9022 ixova	9072 izazi	9122 iznog	9172 jabab	9222 janez	9272 jeaku	9322 jelat	9372 jezib	9422 jinab	9472 jocil
9023 ixozi	9073 izbab	9123 iznux	9173 jabev	9223 janip	9273 jeamo	9323 jelov	9373 jezly	9423 jinev	9473 jocox
9024 ixpew	9074 izbev	9124 iznyk	9174 jabis	9224 janog	9274 jeany	9324 jelul	9374 jezoc	9424 jinig	9474 jocun
9025 ixpha	9075 izbis	9125 izoco	9175 jabuk	9225 janux	9275 jeapt	9325 jelyg	9375 jezuz	9425 jinon	9475 joday
9026 ixpis	9076 izbon	9126 izohm	9176 jacem	9226 janyk	9276 jeava	9326 jemab	9376 jiaci	9426 jinuk	9476 jodef
9027 ixple	9077 izbuk	9127 izoki	9177 jaciz	9227 jaoco	9277 jebaz	9327 jemev	9377 jiado	9427 jinyp	9477 jodoz
9028 ixpoy	9078 izcob	9128 izolt	9178 jacra	9228 jaohm	9278 jebin	9328 jemim	9378 jiafu	9428 jiody	9478 joduv
9029 ixram	9079 izcra	9129 izons	9179 jacyc	9229 jaoki	9279 jebok	9329 jemon	9379 jiaga	9429 jiola	9479 joect
9030 ixrel	**9080** izcyc	**9130** izowe	**9180** jadap	**9230** jaolt	**9280** jebre	**9330** jemuk	**9380** jialk	**9430** jiomo	**9480** joegu
9031 ixroj	9081 izdap	9131 izpan	9181 jadof	9231 jaons	9281 jebus	9331 jenan	9381 jibam	9431 jiong	9481 joeko
9032 ixruw	9082 izdes	9132 izpet	9182 jaduc	9232 jaowe	9282 jecam	9332 jeney	9382 jibel	9432 jiose	9482 joems
9033 ixsax	9083 izdof	9133 izphy	9183 jaeha	9233 japhy	9283 jecel	9333 jenod	9383 jiboy	9433 jipad	9483 joepy
9034 ixske	9084 izduc	9134 izpib	9184 jaelu	9234 japib	9284 jecis	9334 jenum	9384 jibud	9434 jipec	9484 joexa
9035 ixsli	9085 izelu	9135 izpod	9185 jaeny	9235 japod	9285 jecow	9335 jeogi	9385 jicac	9435 jipol	9485 jofah
9036 ixsov	9086 izeny	9136 izpum	9186 jaepo	9236 japum	9286 jecud	9336 jeomy	9386 jiclu	9436 jipub	9486 jofem
9037 ixspu	9087 izets	9137 izrac	9187 jaets	9237 jarac	9287 jecyp	9337 jeono	9387 jicom	9437 jiseb	9487 jofyl
9038 ixsyc	9088 izexi	9138 izreg	9188 jaexi	9238 jareg	9288 jedal	9338 jeork	9388 jicys	9438 jitah	9488 jogal
9039 ixtaf	9089 izeze	9139 izrom	9189 jaeze	9239 jarom	9289 jedik	9339 jeosa	9389 jidan	9439 jitim	9489 jogen
9040 ixtes	**9090** izfay	**9140** izsat	**9190** jafay	**9240** jasat	**9290** jedos	**9340** jeovu	**9390** jidey	**9440** jitob	**9490** jogiv
9041 ixtic	9091 izfex	9141 izsoy	9191 jafex	9241 jasoy	9291 jeduy	9341 jepap	9391 jidif	9441 jityn	9491 jogut
9042 ixtog	9092 izfil	9142 izsul	9192 jafil	9242 jasul	9292 jeeka	9342 jepes	9392 jidod	9442 jiuco	9492 joheh
9043 ixtym	9093 izfoh	9143 iztax	9193 jafoh	9243 jatax	9293 jeelb	9343 jepiz	9393 jidut	9443 jiuda	9493 johin
9044 ixuch	9094 izfuz	9144 izter	9194 jafuz	9244 jater	9294 jeenu	9344 jepro	9394 jiecu	9444 jiuky	9494 johuy
9045 ixudu	9095 izgas	9145 iztid	9195 jagas	9245 jatid	9295 jeesi	9345 jepuc	9395 jiego	9445 jiume	9495 joidu
9046 ixums	9096 izgek	9146 iztup	9196 jagek	9246 jatup	9296 jefad	9346 jerar	9396 jieli	9446 jiuph	9496 joily
9047 ixunk	9097 izgit	9147 iztys	9197 jagit	9247 jatys	9297 jefec	9347 jerit	9397 jierx	9447 jiuvi	9497 joino
9048 ixuri	9098 izgor	9148 izuby	9198 jagor	9248 jauby	9298 jeflo	9348 jeryx	9398 jieve	9448 jiuzu	9498 joiva
9049 ixuto	9099 izgug	9149 izufu	9199 jagug	9249 jaufu	9299 jefry	9349 jesas	9399 jifap	9449 jivay	9499 joize

9500 jojel	9550 jouge	9600 juidy	9650 juveh	9700 kahet	9750 karon	9800 keewo	9850 keorb	9900 kifuf	9950 kiubu
9501 jojus	9551 joulo	9601 juifi	9651 juvip	9701 kahor	9751 karuk	9801 kefan	9851 keoxa	9901 kigaz	9951 kiude
9502 jokan	9552 joups	9602 juips	9652 juvow	9702 kahun	9752 kasad	9802 kefew	9852 kepin	9902 kigli	9952 kiufi
9503 jokey	9553 jouxu	9603 juire	9653 juvus	9703 kahym	9753 kasec	9803 kefod	9853 keplu	9903 kigoc	9953 kivis
9504 jokip	9554 jouza	9604 juita	9654 juvyt	9704 kaici	9754 kasol	9804 kefri	9854 kepyx	9904 kigre	9954 kivon
9505 jokod	9555 jovac	9605 juker	9655 juwav	9705 kaigy	9755 kaspy	9805 kefum	9855 kerad	9905 kigud	9955 kiwaf
9506 jokum	9556 joveg	9606 jukig	9656 juwec	9706 kaiku	9756 kasti	9806 kefys	9856 kerec	9906 kihac	9956 kiwel
9507 jolap	9557 jovom	9607 jukup	9657 juwon	9707 kailo	9757 kasub	9807 kegap	9857 kerol	9907 kihez	9957 kiwip
9508 joles	9558 jowak	9608 jukyf	9658 juxol	9708 kaima	9758 kated	9808 kegiz	9858 kerub	9908 kihid	9958 kiwog
9509 jolik	9559 jowic	9609 julem	9659 juyba	9709 kaise	9759 katip	9809 kegof	9859 kesaf	9909 kihob	9959 kiwye
9510 joluc	9560 jowol	9610 julin	9660 juzid	9710 kajap	9760 katox	9810 keguc	9860 kesed	9910 kijad	9960 kixar
9511 jomar	9561 jozax	9611 julob	9661 juzle	9711 kajes	9761 katuc	9811 kegym	9861 keshy	9911 kijek	9961 kixen
9512 jomez	9562 juafy	9612 juluf	9662 juzot	9712 kajil	9762 katyt	9812 kehak	9862 kesig	9912 kijow	9962 kixim
9513 jomid	9563 juals	9613 julyx	9663 kaacy	9713 kajof	9763 kaubo	9813 kehey	9863 kesoc	9913 kijur	9963 kizap
9514 jomog	9564 juamu	9614 jumal	9664 kaadu	9714 kajux	9764 kauka	9814 kehoz	9864 kesuz	9914 kikaw	9964 kizus
9515 jomux	9565 juank	9615 jumen	9665 kaafa	9715 kakar	9765 kaulu	9815 kehux	9865 ketav	9915 kikem	9965 koabu
9516 jomys	9566 juapi	9616 jumos	9666 kaale	9716 kakib	9766 kaune	9816 keich	9866 ketem	9916 kikot	9966 koaca
9517 jonet	9567 juart	9617 jumyr	9667 kaani	9717 kakog	9767 kavat	9817 keido	9867 ketis	9917 kikuk	9967 koads
9518 jonor	9568 juavo	9618 junam	9668 kaaph	9718 kalas	9768 kavix	9818 keine	9868 ketuf	9918 kilav	9968 koafi
9519 jonug	9569 juawa	9619 junel	9669 kaarx	9719 kalek	9769 kavoc	9819 keisi	9869 keuck	9919 kilep	9969 koago
9520 jonyx	9570 juaxe	9620 junit	9670 kaato	9720 kalim	9770 kawal	9820 keiza	9870 keuno	9920 kilir	9970 koalm
9521 jooga	9571 jubil	9621 junud	9671 kabem	9721 kalow	9771 kawiv	9821 kejar	9871 keuzi	9921 kilok	9971 koane
9522 joolu	9572 juboz	9622 junyc	9672 kabid	9722 kalug	9772 kaxis	9822 kejez	9872 kevag	9922 kilyc	9972 koaty
9523 joopi	9573 jubys	9623 juocu	9673 kabuf	9723 kalyb	9773 kayeb	9823 kejog	9873 kevil	9923 kimat	9973 koawk
9524 joovo	9574 jucan	9624 juoko	9674 kabyx	9724 kamaj	9774 kayom	9824 kekas	9874 kevox	9924 kimey	9974 kobas
9525 jooxe	9575 jucey	9625 juoma	9675 kacaw	9725 kameh	9775 kazag	9825 kekek	9875 kevun	9925 kimup	9975 kobek
9526 jopav	9576 jucod	9626 juope	9676 kacir	9726 kamik	9776 kazoz	9826 kekid	9876 kewir	9926 kinak	9976 kobli
9527 jopex	9577 jucum	9627 juosy	9677 kacoy	9727 kamop	9777 keada	9827 kekor	9877 kewob	9927 kiniz	9977 kobor
9528 jopop	9578 judaw	9628 juoti	9678 kacut	9728 kamur	9778 keame	9828 kekug	9878 keyet	9928 kinof	9978 kobug
9529 jopur	9579 judis	9629 jupas	9679 kadac	9729 kamyd	9779 keans	9829 kekyp	9879 kezic	9929 kinut	9979 kocav
9530 jopyk	9580 judly	9630 jupek	9680 kadeg	9730 kanax	9780 kearg	9830 kelaw	9880 kezuk	9930 kioby	9980 kocit
9531 joraz	9581 judur	9631 jupix	9681 kadin	9731 kaner	9781 keavu	9831 keleh	9881 kiabs	9931 kiocs	9981 kocop
9532 jorep	9582 juelf	9632 jupor	9682 kadyp	9732 kanov	9782 kebal	9832 kelit	9882 kiajo	9932 kiogn	9982 kocur
9533 jorif	9583 juene	9633 jupug	9683 kaefi	9733 kanup	9783 keben	9833 kelur	9883 kiake	9933 kioka	9983 kocyg
9534 jorok	9584 jueva	9634 jurap	9684 kaega	9734 kanyg	9784 kebix	9834 kelyd	9884 kialu	9934 kiole	9984 kodax
9535 jorud	9585 juexu	9635 jures	9685 kaeke	9735 kaoch	9785 kebos	9835 kemax	9885 kiarc	9935 kioro	9985 kodiz
9536 joryc	9586 jufar	9636 juriv	9686 kaelm	9736 kaode	9786 kecli	9836 kemer	9886 kiavi	9936 kiotz	9986 kodov
9537 josab	9587 jufez	9637 jurof	9687 kaepu	9737 kaoja	9787 kecre	9837 kemib	9887 kicah	9937 kioxi	9987 kodup
9538 josev	9588 jufik	9638 juruc	9688 kaery	9738 kaomi	9788 kedam	9838 kemov	9888 kicig	9938 kipaj	9988 koemi
9539 josis	9589 jufog	9639 juryb	9689 kaeth	9739 kaops	9789 kedel	9839 kemut	9889 kidet	9939 kipho	9989 koepa
9540 joson	9590 jufux	9640 jusac	9690 kafam	9740 kaorf	9790 kediv	9840 kemyn	9890 kidom	9940 kipuy	9990 koero
9541 josup	9591 jugaf	9641 juseg	9691 kafel	9741 kaotu	9791 kedro	9841 kenaz	9891 kidri	9941 kiray	9991 koest
9542 josym	9592 juged	9642 jusir	9692 kafot	9742 kapaz	9792 kedud	9842 kenep	9892 kidyl	9942 kiroh	9992 koewe
9543 jotad	9593 jugno	9643 jusom	9693 kafud	9743 kapep	9793 kedyt	9843 kenok	9893 kiebi	9943 kisib	9993 kofaz
9544 jotec	9594 jugry	9644 jusuk	9694 kagan	9744 kapit	9794 keebe	9844 kenus	9894 kieft	9944 kisko	9994 kofep
9545 jotir	9595 juguz	9645 jusyn	9695 kagew	9745 kapok	9795 keecs	9845 keofo	9895 kielk	9945 kisla	9995 kofin
9546 jotub	9596 juhax	9646 jutak	9696 kagic	9746 kapus	9796 keeju	9846 keoke	9896 kiena	9946 kiswe	9996 kofly
9547 jotyp	9597 juhic	9647 jutet	9697 kagod	9747 karab	9797 keend	9847 keold	9897 kierp	9947 kisyf	9997 kofok
9548 joubi	9598 juhub	9648 jutoc	9698 kagum	9748 karev	9798 keerf	9848 keont	9898 kifol	9948 kitan	9998 kofus
9549 joucy	9599 juick	9649 jutun	9699 kahay	9749 kariz	9799 keeta	9849 keopu	9899 kifra	9949 kitux	9999 kogab

kogev kogis kogon kogyr

[redacted]

SECTION B.

SECTION B

0000 kohew	**0050** kosyk	**0100** kugly	**0150** kurav	**0200** kyele	**0250** kyowl	**0300** ladis	**0350** lapix	**0400** lecad	**0450** lelud
0001 kohof	0051 kotat	0101 kugob	0151 kurex	0201 kyens	0251 kypam	0301 ladly	0351 lapor	0401 leceo	0451 lemex
0002 kohuc	0052 kotez	0102 kuguf	0152 kurip	0202 kyeph	0252 kypel	0302 ladur	0352 lapug	0402 lecol	0452 lemig
0003 kohyd	0053 kotoy	0103 kuhil	0153 kurot	0203 kyerd	0253 kypiv	0303 laejo	0353 larap	0403 lecri	0453 lemum
0004 koibs	0054 kotru	0104 kuhow	0154 kurul	0204 kyeto	0254 kypot	0304 laelf	0354 lariv	0404 lecub	0454 lemyl
0005 koice	0055 koufo	0105 kuibu	0155 kurym	0205 kyewi	0255 kypud	0305 laene	0355 laruc	0405 lecyn	0455 lenap
0006 koipi	0056 kouma	0106 kuify	0156 kusar	0206 kyfac	0256 kyran	0306 lafar	0356 laryb	0406 ledaf	0456 lenes
0007 koirl	0057 kound	0107 kuige	0157 kusez	0207 kyfeg	0257 kyrew	0307 lafez	0357 lasir	0407 leded	0457 lenic
0008 kojac	0058 kouse	0108 kuind	0158 kusin	0208 kyfib	0258 kyrik	0308 lafik	0358 lasom	0408 ledil	0458 lenot
0009 kojeg	0059 kouti	0109 kuirk	0159 kusog	0209 kyfom	0259 kyrod	0309 lafog	0359 lasuk	0409 ledoc	0459 leobu
0010 kojuw	**0060** kovan	**0110** kuisl	**0160** kutaz	**0210** kyfru	**0260** kyrum	**0310** lafux	**0360** lasyn	**0410** leduz	**0460** leogy
0011 kokad	0061 kovik	0111 kuivo	0161 kutus	0211 kygad	0261 kyryl	0311 lagaf	0361 latak	0411 ledyx	0461 leols
0012 kokec	0062 kovum	0112 kuixa	0162 kutyr	0212 kygec	0262 kysap	0312 laged	0362 latet	0412 leeba	0462 leond
0013 kokir	0063 kowar	0113 kujal	0163 kuvas	0213 kygif	0263 kyses	0313 lagry	0363 latij	0413 leedo	0463 leorc
0014 kokol	0064 kowes	0114 kujen	0164 kuvej	0214 kygol	0264 kysiz	0314 laguz	0364 latoc	0414 leege	0464 leoze
0015 kokub	0065 kowot	0115 kujos	0165 kuvib	0215 kygub	0265 kysof	0315 lahax	0365 latun	0415 leeki	0465 lepar
0016 kolaf	0066 koyaw	0116 kujut	0166 kuvor	0216 kyhav	0266 kysuc	0316 lahic	0366 latyd	0416 leemp	0466 lepez
0017 koled	0067 kozey	0117 kukam	0167 kuwah	0217 kyhit	0267 kytaw	0317 lahub	0367 laugi	0417 leesu	0467 lepog
0018 kolip	0068 kozog	0118 kukel	0168 kuwey	0218 kyhoy	0268 kytex	0318 laick	0368 launu	0418 lefag	0468 lepyt
0019 koloc	0069 kozul	0119 kukif	0169 kuwho	0219 kyhul	0269 kythi	0319 laidy	0369 laupo	0419 lefeb	0469 leras
0020 koluz	**0070** kuagy	**0120** kukro	**0170** kuyon	**0220** kyibe	**0270** kytor	**0320** laifi	**0370** laurb	**0420** lefir	**0470** lerek
0021 kolyn	0071 kuano	0121 kukud	0171 kuyra	0221 kyica	0271 kyuga	0321 laips	0371 lausa	0421 lefox	0471 lerib
0022 komag	0072 kuare	0122 kulan	0172 kuyug	0222 kyifo	0272 kyumi	0322 laire	0372 laute	0422 lefun	0472 leror
0023 komeb	0073 kuasi	0123 kulet	0173 kuzab	0223 kyigu	0273 kyurn	0323 laita	0373 laveh	0423 legef	0473 lerug
0024 komil	0074 kuatu	0124 kulod	0174 kuzef	0224 kyimp	0274 kyutu	0324 lajul	0374 lavip	0424 legid	0474 lesav
0025 komox	0075 kuawl	0125 kulum	0175 kuzit	0225 kyits	0275 kyvar	0325 lakig	0375 lavow	0425 legoz	0475 lesim
0026 komun	0076 kubad	0126 kumap	0176 kyaba	0226 kykag	0276 kyvez	0326 lakup	0376 lavus	0426 legul	0476 lesly
0027 komyc	0077 kubec	0127 kumes	0177 kyach	0227 kykeb	0277 kyvin	0327 lakyf	0377 lavyt	0427 legyb	0477 lesur
0028 konah	0078 kubol	0128 kumiz	0178 kyaim	0228 kykil	0278 kyvog	0328 lalem	0378 lawav	0428 lehaw	0478 lethy
0029 konef	0079 kubub	0129 kumof	0179 kyaju	0229 kykun	0279 kywat	0329 lalin	0379 lawec	0429 lehok	0479 letit
0030 konib	**0080** kucaf	**0130** kumuc	**0180** kyaly	**0230** kylay	**0280** kywid	**0330** lalob	**0380** lawon	**0430** lehut	**0480** letow
0031 konoz	0081 kuced	0131 kumyt	0181 kyant	0231 kylef	0281 laafy	0331 laluf	0381 laxol	0431 lehyc	0481 letre
0032 konyl	0082 kuchi	0132 kunac	0182 kyase	0232 kylig	0282 laals	0332 lalyx	0382 layba	0432 leibi	0482 leuja
0033 kooba	0083 kucuz	0133 kuneg	0183 kyazo	0233 kyloz	0283 laamu	0333 lamal	0383 lazid	0433 leicu	0483 leult
0034 kooci	0084 kucyk	0134 kuniv	0184 kybax	0234 kyluv	0284 laank	0334 lamen	0384 lazle	0434 leida	0484 leunz
0035 kooky	0085 kudag	0135 kunom	0185 kyber	0235 kymak	0285 laapi	0335 lamos	0385 lazot	0435 leiko	0485 levov
0036 kooxo	0086 kudeb	0136 kunys	0186 kyblo	0236 kymem	0286 laart	0336 lamyr	0386 leabo	0436 leing	0486 levup
0037 kopak	0087 kudid	0137 kuoce	0187 kybup	0237 kymip	0287 laavo	0337 lanam	0387 leacs	0437 leisy	0487 lewaz
0038 kopem	0088 kudox	0138 kuodo	0188 kybyn	0238 kymob	0288 laawa	0338 lanel	0388 leald	0438 lejak	0488 lewet
0039 koplo	0089 kudun	0139 kuoli	0189 kycaz	0239 kymuf	0289 laaxe	0339 lanit	0389 leapy	0439 lejem	0489 lewod
0040 kopry	**0090** kueky	**0140** kuoms	**0190** kycep	**0240** kymyg	**0290** labaj	**0340** lanud	**0390** learm	**0440** lejob	**0490** leyew
0041 kopuf	0091 kueld	0141 kuonz	0191 kycro	0241 kynal	0291 labil	0341 lanyc	0391 leate	0441 lejuf	0491 leyla
0042 koral	0092 kuepi	0142 kuory	0192 kycus	0242 kynek	0292 laboz	0342 laocu	0392 leawn	0442 lekal	0492 leyus
0043 koros	0093 kueru	0143 kuoth	0193 kydev	0243 kynix	0293 labys	0343 laoko	0393 leaxi	0443 leken	0493 lezah
0044 korut	0094 kuete	0144 kuowa	0194 kydic	0244 kynyo	0294 lacan	0344 laoma	0394 leaza	0444 lekos	0494 lezin
0045 kosam	0095 kufat	0145 kuoxu	0195 kydon	0245 kyobs	0295 lacey	0345 laope	0395 lebac	0445 lekyk	0495 liaha
0046 kosel	0096 kufeh	0146 kupax	0196 kydry	0246 kyoda	0296 lacod	0346 laosy	0396 lebeg	0446 lelam	0496 lialp
0047 kosho	0097 kufoz	0147 kupic	0197 kyduk	0247 kyofi	0297 lacum	0347 laoti	0397 lebif	0447 lelel	0497 liamb
0048 kosif	0098 kugak	0148 kupov	0198 kyebu	0248 kyojo	0298 ladaw	0348 lapas	0398 leblu	0448 lelix	0498 liark
0049 kosud	0099 kugem	0149 kupup	0199 kyefa	0249 kyoku	0299 ladex	0349 lapek	0399 lebro	0449 lelon	0499 liayo

0500 liboj	**0550** lispa	**0600** loete	**0650** loowa	**0700** ludin	**0750** lunax	**0800** lybyc	**0850** lymif	**0900** mabyl	**0950** manav
0501 licat	0551 lisuh	0601 lofat	0651 looxu	0701 ludul	0751 lunif	0801 lycap	0851 lymom	0901 macay	0951 manop
0502 licen	0552 litel	0602 lofeh	0652 lopic	0702 ludyp	0752 lunov	0802 lyces	0852 lynew	0902 macet	0952 manur
0503 liclo	0553 litif	0603 lofoz	0653 lorav	0703 luefi	0753 lunup	0803 lyciv	0853 lynik	0903 macoz	0953 maobe
0504 licuf	0554 litom	0604 logak	0654 lorex	0704 luega	0754 lunyg	0804 lycof	0854 lynol	0904 madak	0954 maock
0505 licyr	0555 litug	0605 logem	0655 lorip	0705 lueke	0755 luoch	0805 lycuc	0855 lyobo	0905 madem	0955 maolo
0506 lidew	0556 lityb	0606 logly	0656 lorot	0706 luelm	0756 luode	0806 lycyx	0856 lyofy	0906 madob	0956 maomu
0507 lidir	0557 liufy	0607 logob	0657 lorul	0707 luepu	0757 luoja	0807 lydar	0857 lyome	0907 maduf	0957 maoni
0508 lidot	0558 liula	0608 loguf	0658 lorym	0708 luery	0758 luomi	0808 lydez	0858 lyonc	0908 madyn	0958 maopy
0509 lidum	0559 liumo	0609 lohil	0659 losez	0709 lueth	0759 luops	0809 lydip	0859 lyots	0909 maebo	0959 maota
0510 lidys	**0560** liupi	**0610** lohow	**0660** losog	**0710** lufam	**0760** luorf	**0810** lydog	**0860** lypaf	**0910** maede	**0960** mapev
0511 lieco	0561 liuvu	0611 loibu	0661 losux	0711 lufel	0761 luotu	0811 lydux	0861 lyped	0911 maefy	0961 mapif
0512 liers	0562 livep	0612 loify	0662 lotyr	0712 lufig	0762 lupaz	0812 lyedy	0862 lypil	0912 maeku	0962 mapon
0513 lifaf	0563 livod	0613 loige	0663 loudi	0713 lufot	0763 lupep	0813 lyefo	0863 lypoc	0913 maela	0963 mapuk
0514 lifli	0564 liwad	0614 loind	0664 lougu	0714 lufud	0764 lupit	0814 lyelv	0864 lypuz	0914 maern	0964 mapyc
0515 lifos	0565 liweg	0615 loirk	0665 louko	0715 lugan	0765 lupok	0815 lyemu	0865 lyrag	0915 maeti	0965 marov
0516 lifre	0566 liwhy	0616 loisl	0666 loule	0716 lugew	0766 lurab	0816 lyerg	0866 lyreb	0916 mafaw	0966 maryd
0517 liglu	0567 liwim	0617 loivo	0667 louna	0717 lugic	0767 lurev	0817 lyese	0867 lyrid	0917 mafen	0967 masaz
0518 liham	0568 liwov	0618 loixa	0668 lovas	0718 lugod	0768 luriz	0818 lyfas	0868 lyrox	0918 mafic	0968 masep
0519 liheb	0569 lixan	0619 lojal	0669 lovej	0719 lugum	0769 luruk	0819 lyfix	0869 lyrun	0919 mafoy	0969 masmi
0520 lihud	**0570** lixor	**0620** lojen	**0670** lovib	**0720** luhay	**0770** luspy	**0820** lyfor	**0870** lysaw	**0920** mafut	**0970** masow
0521 lijok	0571 lizak	0621 lojos	0671 lovor	0721 luhet	0771 lusti	0821 lyfug	0871 lysef	0921 magam	0971 masus
0522 lijun	0572 lizem	0622 lojut	0672 lowah	0722 luhor	0772 luted	0822 lygav	0872 lysit	0922 magel	0972 matal
0523 likah	0573 liziz	0623 lokam	0673 lowey	0723 luhun	0773 lutip	0823 lygim	0873 lysoz	0923 magip	0973 mateg
0524 liket	0574 lizob	0624 lokel	0674 lowho	0724 luhym	0774 lutox	0824 lygop	0874 lysuv	0924 magru	0974 matum
0525 likow	0575 lizup	0625 lokif	0675 loyon	0725 luici	0775 luvat	0825 lygur	0875 lysys	0925 mahaf	0975 matyx
0526 likri	0576 loagy	0626 lokro	0676 loyra	0726 luigy	0776 luvix	0826 lyhab	0876 lyten	0926 maher	0976 maucu
0527 likyx	0577 loala	0627 lokud	0677 loyug	0727 luiku	0777 luvoc	0827 lyhek	0877 lytha	0927 mahis	0977 mauki
0528 lilaz	0578 loano	0628 lolan	0678 lozab	0728 luilo	0778 luwal	0828 lyhig	0878 lytry	0928 mahod	0978 maung
0529 lilic	0579 loare	0629 lolet	0679 lozef	0729 luima	0779 luwiv	0829 lyhon	0879 lytuk	0929 mahuz	0979 maupa
0530 lilus	**0580** loasi	**0630** lolis	**0680** lozit	**0730** lujap	**0780** luxis	**0830** lyhup	**0880** lyuba	**0930** maidi	**0980** mauso
0531 lilyl	0581 loatu	0631 lolod	0681 luacy	0731 lujes	0781 luyeb	0831 lyimo	0881 lyufe	0931 maifu	0981 mavad
0532 limay	0582 loawl	0632 lolum	0682 luadu	0732 lujil	0782 luyom	0832 lyiny	0882 lyuni	0932 mailk	0982 mavec
0533 limis	0583 lobec	0633 lomap	0683 luafa	0733 lujof	0783 luzag	0833 lyipu	0883 lyvak	0933 maisa	0983 mavir
0534 limul	0584 lobol	0634 lomes	0684 luale	0734 lujux	0784 luzoz	0834 lyiti	0884 lyvem	0934 maivy	0984 mavol
0535 linex	0585 lobub	0635 lomiz	0685 luani	0735 lukar	0785 lyabe	0835 lyiwa	0885 lyvic	0935 majan	0985 mawac
0536 linil	0586 locaf	0636 lomof	0686 luaph	0736 lukib	0786 lyack	0836 lykaz	0886 lyvul	0936 majeb	0986 mawed
0537 linoy	0587 loced	0637 lomuc	0687 luarx	0737 lukog	0787 lyaft	0837 lykep	0887 lyweh	0937 majik	0987 mayoc
0538 linub	0588 lochi	0638 lomyt	0688 luato	0738 lulas	0788 lyagi	0838 lykin	0888 maaby	0938 makap	0988 mazat
0539 liofa	0589 lococ	0639 lonac	0689 lubak	0739 lulek	0789 lyahu	0839 lykok	0889 maadz	0939 makes	0989 mazom
0540 liolm	**0590** locuz	**0640** loneg	**0690** lubem	**0740** lulim	**0790** lyaja	**0840** lykus	**0890** maaho	**0940** makim	**0990** meagy
0541 liopo	0591 locyk	0641 loniv	0691 lubid	0741 lulow	0791 lyamy	0841 lykym	0891 maalc	0941 makuc	0991 meala
0542 liorn	0592 lodag	0642 lonom	0692 lubob	0742 lulug	0792 lyarn	0842 lylal	0892 maana	0942 malez	0992 meano
0543 liozu	0593 lodeb	0643 lonys	0693 lubuf	0743 lulyb	0793 lyasp	0843 lylex	0893 maari	0943 malid	0993 measi
0544 liply	0594 lodid	0644 looce	0694 lubyx	0744 lumaj	0794 lyawo	0844 lylib	0894 maasm	0944 malog	0994 meatu
0545 lipru	0595 lodox	0645 loodo	0695 lucaw	0745 lumeh	0795 lyban	0845 lyloh	0895 maaxu	0945 malux	0995 mebad
0546 liraj	0596 lodun	0646 looli	0696 lucoy	0746 lumik	0796 lybet	0846 lylut	0896 mabag	0946 malym	0996 mebol
0547 lirey	0597 loeky	0647 loonz	0697 lucut	0747 lumop	0797 lybiz	0847 lylyr	0897 mabef	0947 mamek	0997 mecaf
0548 lisid	0598 loeld	0648 loory	0698 ludac	0748 lumur	0798 lybod	0848 lymac	0898 mabib	0948 mamit	0998 meced
0549 lisky	0599 loepi	0649 looth	0699 ludeg	0749 lumyd	0799 lybum	0849 lymeg	0899 mabun	0949 mamug	0999 mechi

mecoc mecuz medeb medid

SECTION B

1000 medox
1001 meeky
1002 meeru
1003 meete
1004 mefeh
1005 mefim
1006 megak
1007 megly
1008 meguf
1009 meibu
1010 meify
1011 meige
1012 meind
1013 meirk
1014 meisl
1015 meivo
1016 mejal
1017 mejos
1018 mekel
1019 mekif
1020 mekro
1021 mekud
1022 melan
1023 melod
1024 memes
1025 memiz
1026 memof
1027 memuc
1028 memyt
1029 menac
1030 meneg
1031 meniv
1032 menys
1033 meoce
1034 meodo
1035 meoli
1036 meoms
1037 meonz
1038 meory
1039 meoth
1040 meowa
1041 meoxu
1042 meper
1043 mepov
1044 merav
1045 merul
1046 merym
1047 mesez
1048 mesog
1049 mesux
1050 metaz
1051 metix
1052 metyr
1053 meudi
1054 meugu
1055 meuko
1056 meuna
1057 mevas
1058 mevej
1059 mevib
1060 mevor
1061 mewah
1062 meyra
1063 mezef
1064 miade
1065 miaku
1066 miamo
1067 miany
1068 miapt
1069 miars
1070 miava
1071 mibaz
1072 mibin
1073 mibok
1074 mibre
1075 mibus
1076 mibyt
1077 micam
1078 micel
1079 micud
1080 micyp
1081 midal
1082 miden
1083 miduy
1084 mieka
1085 mielb
1086 mienu
1087 miesi
1088 mifad
1089 mifec
1090 miflo
1091 mifry
1092 mifub
1093 migag
1094 migir
1095 migle
1096 migun
1097 mihah
1098 miheg
1099 mihol
1100 mihur
1101 mijay
1102 mijef
1103 mijid
1104 mijot
1105 mijup
1106 mikak
1107 mikeh
1108 mikob
1109 mikuf
1110 mikys
1111 milat
1112 milov
1113 milul
1114 milyg
1115 mimab
1116 mimev
1117 mimon
1118 mimuk
1119 minan
1120 miney
1121 minod
1122 miogi
1123 miomy
1124 miono
1125 miork
1126 miosa
1127 miovu
1128 mipap
1129 mipes
1130 mipiz
1131 mipro
1132 mipuc
1133 mipyd
1134 mirar
1135 mirit
1136 miryx
1137 mispi
1138 misyl
1139 mitac
1140 mitof
1141 miugo
1142 miuli
1143 miumu
1144 miush
1145 miuta
1146 mivav
1147 mivet
1148 mivif
1149 mivoy
1150 miwax
1151 miwig
1152 miwom
1153 miwyn
1154 mizaf
1155 mized
1156 mizib
1157 mizly
1158 mizoc
1159 mizuz
1160 moanz
1161 moapa
1162 moask
1163 moath
1164 moaxo
1165 mobez
1166 mobly
1167 mocin
1168 mocor
1169 mocug
1170 mocyb
1171 modav
1172 modej
1173 modop
1174 moebs
1175 moegi
1176 moema
1177 moeno
1178 moepe
1179 moert
1180 moezu
1181 mogep
1182 mogil
1183 mogus
1184 mohaz
1185 mohyp
1186 moili
1187 moimu
1188 moina
1189 moipo
1190 mojab
1191 mojif
1192 mojon
1193 mokac
1194 mokyn
1195 molec
1196 moliv
1197 molol
1198 molub
1199 momaf
1200 momed
1201 momoc
1202 momuz
1203 monox
1204 moodu
1205 moogo
1206 moove
1207 mooys
1208 mopir
1209 mopoz
1210 morak
1211 morem
1212 morob
1213 moruf
1214 moscu
1215 moski
1216 motiz
1217 motur
1218 motyl
1219 movam
1220 movel
1221 movij
1222 movot
1223 moyat
1224 mozeh
1225 mozof
1226 muahs
1227 mualb
1228 muamt
1229 muawe
1230 muaya
1231 mubap
1232 mubes
1233 mubig
1234 mubof
1235 mubuc
1236 mubyb
1237 muchy
1238 mucic
1239 mudas
1240 mudek
1241 mudor
1242 mudug
1243 mudyc
1244 muelz
1245 muemy
1246 muept
1247 muere
1248 mueso
1249 muevu
1250 mufav
1251 mufey
1252 mufiz
1253 mufop
1254 mufur
1255 mugib
1256 mugov
1257 muhaj
1258 muhir
1259 muhox
1260 muhuk
1261 muhys
1262 muics
1263 muigo
1264 muika
1265 muiph
1266 muity
1267 muizu
1268 mujat
1269 mujew
1270 mukab
1271 mukev
1272 mukis
1273 muklu
1274 mukon
1275 mukyr
1276 mulac
1277 muleg
1278 mulif
1279 mulom
1280 mulyp
1281 mumec
1282 mumix
1283 mumol
1284 munaf
1285 muned
1286 munoc
1287 munuz
1288 muoca
1289 muodi
1290 muoho
1291 muonu
1292 muost
1293 muoxy
1294 mupag
1295 mupeb
1296 mupim
1297 mupow
1298 mupun
1299 muraw
1300 muref
1301 muroz
1302 musah
1303 musem
1304 musip
1305 musob
1306 musuf
1307 mutam
1308 mutep
1309 mutiv
1310 mutot
1311 mutul
1312 muval
1313 muven
1314 muvit
1315 muvos
1316 muway
1317 muwex
1318 muxin
1319 muzan
1320 muzik
1321 muzum
1322 myabo
1323 myacs
1324 myald
1325 myapy
1326 myaxi
1327 myaza
1328 mybif
1329 myblu
1330 mybro
1331 mycec
1332 mydaf
1333 myded
1334 mydil
1335 mydoc
1336 myduz
1337 mydyx
1338 myeba
1339 myech
1340 myedo
1341 myege
1342 myemp
1343 myesu
1344 myfeb
1345 mygat
1346 mygef
1347 mygid
1348 mygoz
1349 mygul
1350 myhok
1351 myhyc
1352 myibi
1353 myicu
1354 myiko
1355 mying
1356 myith
1357 mykal
1358 mykyk
1359 mylam
1360 mylel
1361 mylix
1362 mylon
1363 mylud
1364 mymex
1365 mymig
1366 mymoh
1367 mymyl
1368 mynes
1369 mynic
1370 myobu
1371 myogy
1372 myols
1373 myond
1374 myoto
1375 myoze
1376 mypar
1377 mypez
1378 mypik
1379 mypog
1380 myras
1381 myrek
1382 myror
1383 myrug
1384 mysav
1385 mysim
1386 mysur
1387 mytab
1388 mytre
1389 myube
1390 myuja
1391 myult
1392 myumy
1393 myunz
1394 myvov
1395 myvup
1396 mywaz
1397 mywod
1398 naagy
1399 naala
1400 naano
1401 naasi
1402 naatu
1403 naawl
1404 nabec
1405 nabol
1406 nabub
1407 nacaf
1408 naced
1409 nachi
1410 nacoc
1411 nacuz
1412 nacyk
1413 nadag
1414 nadeb
1415 nadid
1416 nadox
1417 nadun
1418 naeky
1419 naeld
1420 naepi
1421 naeru
1422 naete
1423 nafat
1424 nafeh
1425 nafim
1426 nafoz
1427 nagak
1428 nagem
1429 nagly
1430 nagob
1431 naguf
1432 nahil
1433 nahow
1434 naibu
1435 naify
1436 naige
1437 naind
1438 nairk
1439 naisl
1440 naivo
1441 naixa
1442 najal
1443 najen
1444 najos
1445 najut
1446 nakam
1447 nakel
1448 nakif
1449 nakro
1450 nakud
1451 nalan
1452 nalet
1453 nalis
1454 nalod
1455 nalum
1456 namap
1457 namiz
1458 namof
1459 namuc
1460 namyt
1461 nanac
1462 naneg
1463 naniv
1464 nanys
1465 naoce
1466 naodo
1467 naoli
1468 naoms
1469 naonz
1470 naory
1471 naoth
1472 naowa
1473 naoxu
1474 naper
1475 napic
1476 napov
1477 narav
1478 narex
1479 narip
1480 narot
1481 narul
1482 narym
1483 nasar
1484 nasez
1485 nasin
1486 nasog
1487 nasux
1488 nataz
1489 natyr
1490 naudi
1491 naugu
1492 nauko
1493 nauna
1494 navej
1495 navib
1496 nawah
1497 nawey
1498 nawho
1499 nayon

1500 nayra	**1550** neipu	**1600** neuzo	**1650** nilem	**1700** niwec	**1750** nokak	**1800** noyem	**1850** nuich	**1900** nusoc	**1950** nygeg
1501 nayug	1551 neiti	1601 nevak	1651 nilin	1701 niwon	1751 nokeh	1801 noyux	1851 nuido	1901 nusuz	1951 nygin
1502 nazab	1552 neiwa	1602 nevem	1652 nilob	1702 nixol	1752 nokob	1802 nozaf	1852 nuine	1902 nutav	1952 nygom
1503 nazef	1553 nejah	1603 nevic	1653 niluf	1703 nizid	1753 nokuf	1803 nozed	1853 nuisi	1903 nutem	1953 nygyl
1504 neabe	1554 nejis	1604 nevul	1654 nilyx	1704 nizle	1754 nokys	1804 nozib	1854 nuiza	1904 nutis	1954 nyhad
1505 neack	1555 nekaz	1605 neyle	1655 nimal	1705 nizot	1755 nolat	1805 nozly	1855 nujar	1905 nutuf	1955 nyhif
1506 neaft	1556 nekep	1606 neyot	1656 nimen	1706 noade	1756 nolov	1806 nozoc	1856 nujez	1906 nuvag	1956 nyhot
1507 neagi	1557 nekin	1607 neyub	1657 nimyr	1707 noaku	1757 nolul	1807 nozuz	1857 nujog	1907 nuvil	1957 nyhug
1508 neahu	1558 nekok	1608 nezoy	1658 ninam	1708 noamo	1758 nolyg	1808 nuaco	1858 nukas	1908 nuvox	1958 nyhyr
1509 neaja	1559 nekus	1609 niafy	1659 ninel	1709 noany	1759 nomev	1809 nuada	1859 nukek	1909 nuvun	1959 nyicy
1510 neamy	**1560** nekym	**1610** nials	**1660** ninit	**1710** noapt	**1760** nomon	**1810** nuame	**1860** nukid	**1910** nuwir	**1960** nyigi
1511 nearn	1561 nelal	1611 niamu	1661 ninud	1711 noars	1761 nomuk	1811 nuans	1861 nukor	1911 nuwob	1961 nyinz
1512 neasp	1562 nelex	1612 niank	1662 ninyc	1712 noava	1762 nonan	1812 nuarg	1862 nukug	1912 nuyet	1962 nyiro
1513 neawo	1563 nelib	1613 niapi	1663 niocu	1713 nobaz	1763 noney	1813 nuavu	1863 nukyp	1913 nuyup	1963 nyism
1514 neban	1564 neloh	1614 niart	1664 nioko	1714 nobok	1764 nonod	1814 nubal	1864 nulaw	1914 nuzic	1964 nykaf
1515 nebet	1565 nelut	1615 niavo	1665 nioma	1715 nobre	1765 nonum	1815 nuben	1865 nuleh	1915 nuzuk	1965 nyked
1516 nebiz	1566 nelyr	1616 niawa	1666 niope	1716 nobyt	1766 noogi	1816 nubix	1866 nulit	1916 nyabi	1966 nykli
1517 nebod	1567 nemac	1617 niaxe	1667 niosy	1717 nocam	1767 noomy	1817 nucab	1867 nulur	1917 nyact	1967 nykoc
1518 nebum	1568 nemeg	1618 nibaj	1668 nipas	1718 nocel	1768 noono	1818 nucli	1868 nulyd	1918 nyady	1968 nylag
1519 nebyc	1569 nemif	1619 niboz	1669 nipek	1719 nocow	1769 noork	1819 nucon	1869 numax	1919 nyage	1969 nyleb
1520 necap	**1570** nemom	**1620** nibys	**1670** nipix	**1720** nocud	**1770** noovu	**1820** nucre	**1870** numer	**1920** nyama	**1970** nyliz
1521 neces	1571 nenew	1621 nican	1671 nipor	1721 nocyp	1771 nopap	1821 nudam	1871 numib	1921 nyapo	1971 nylox
1522 neciv	1572 nenik	1622 nicey	1672 nipug	1722 noden	1772 nopes	1822 nudiv	1872 numov	1922 nyaux	1972 nylun
1523 necof	1573 nenol	1623 nicod	1673 nirap	1723 noduy	1773 nopiz	1823 nudro	1873 numut	1923 nybav	1973 nylyt
1524 necuc	1574 neobo	1624 nicum	1674 nires	1724 noecy	1774 nopro	1824 nudud	1874 numyn	1924 nybij	1974 nymah
1525 necyx	1575 neofy	1625 nidaw	1675 niriv	1725 noeka	1775 nopuc	1825 nudyt	1875 nunaz	1925 nyble	1975 nymef
1526 nedar	1576 neome	1626 nidis	1676 nirof	1726 noelb	1776 nopyd	1826 nuebe	1876 nunep	1926 nyboh	1976 nymic
1527 nedez	1577 neonc	1627 nidly	1677 niruc	1727 noenu	1777 norar	1827 nuecs	1877 nunok	1927 nybur	1977 nymoz
1528 nedip	1578 neopa	1628 nidur	1678 nisac	1728 noesi	1778 noryx	1828 nueju	1878 nunus	1928 nycax	1978 nynay
1529 nedog	1579 neots	1629 niejo	1679 niseg	1729 nofad	1779 nosas	1829 nuend	1879 nuofo	1929 nycib	1979 nynem
1530 nedux	**1580** nepil	**1630** nielf	**1680** nisom	**1730** nofec	**1780** nosek	**1830** nuerf	**1880** nuoke	**1930** nycov	**1980** nynid
1531 neefo	1581 nerag	1631 niene	1681 nisuk	1731 noflo	1781 nosor	1831 nueta	1881 nuold	1931 nycup	1981 nynob
1532 neelv	1582 nereb	1632 niexu	1682 nisyn	1732 nofry	1782 nospi	1832 nuewo	1882 nuont	1932 nydaz	1982 nynuf
1533 neemu	1583 nerid	1633 nifar	1683 nitak	1733 nofub	1783 nosug	1833 nufan	1883 nuopu	1933 nydep	1983 nyoak
1534 neerg	1584 nerox	1634 nifez	1684 nitij	1734 nogag	1784 nosyl	1834 nufew	1884 nuorb	1934 nydix	1984 nyofu
1535 neese	1585 nerun	1635 nifik	1685 nitoc	1735 nogir	1785 notac	1835 nufod	1885 nuoxa	1935 nydok	1985 nyomp
1536 nefix	1586 neryf	1636 nifog	1686 nitun	1736 nogle	1786 notil	1836 nufri	1886 nupac	1936 nydus	1986 nyone
1537 nefor	1587 nesaw	1637 nifux	1687 nityd	1737 nogun	1787 notof	1837 nufum	1887 nupeg	1937 nyedi	1987 nyort
1538 nefug	1588 nesef	1638 nigaf	1688 niugi	1738 nohah	1788 nougo	1838 nufys	1888 nupin	1938 nyemo	1988 nyosi
1539 negav	1589 nesit	1639 niged	1689 niunu	1739 noheg	1789 nouli	1839 nugap	1889 nuplu	1939 nyeng	1989 nyowo
1540 negim	**1590** nesoz	**1640** nigno	**1690** niupo	**1740** noiby	**1790** noumu	**1840** nuges	**1890** nupom	**1940** nyerl	**1990** nypal
1541 negop	1591 nesuv	1641 nigry	1691 niurb	1741 noico	1791 noush	1841 nugiz	1891 nupyx	1941 nyesy	1991 nypen
1542 negur	1592 nesys	1642 niguz	1692 niusa	1742 noifa	1792 nouta	1842 nugof	1892 nurad	1942 nyewa	1992 nypig
1543 negyn	1593 neten	1643 nihax	1693 niute	1743 noitu	1793 novav	1843 nuguc	1893 nurec	1943 nyexe	1993 nypos
1544 nehab	1594 netha	1644 nihub	1694 niveh	1744 noive	1794 novet	1844 nugym	1894 nurol	1944 nyfab	1994 nyrat
1545 nehig	1595 netry	1645 nijul	1695 nivip	1745 noizi	1795 novif	1845 nuhak	1895 nurub	1945 nyfev	1995 nyrez
1546 nehon	1596 netuk	1646 niker	1696 nivow	1746 nojay	1796 novoy	1846 nuhey	1896 nusaf	1946 nyfip	1996 nyrhi
1547 nehup	1597 neuba	1647 nikig	1697 nivus	1747 nojef	1797 nowom	1847 nuhip	1897 nused	1947 nyflu	1997 nyrow
1548 neimo	1598 neufe	1648 nikup	1698 nivyt	1748 nojot	1798 nowyn	1848 nuhoz	1898 nushy	1948 nyfon	1998 nyryn
1549 neiny	1599 neuni	1649 nikyf	1699 niwav	1749 nojup	1799 noyca	1849 nuhux	1899 nusig	1949 nygac	1999 nysew

nysil nysod nysum nytas

SECTION B

2000 nytey
2001 nytik
2002 nytol
2003 nytut
2004 nyufa
2005 nyunc
2006 nyurg
2007 nyust
2008 nyuxo
2009 nyuze
2010 nyvap
2011 nyves
2012 nyviv
2013 nywit
2014 nywor
2015 nyxel
2016 obabe
2017 oback
2018 obaft
2019 obagi
2020 obahu
2021 obaja
2022 obamy
2023 obarn
2024 obasp
2025 obawo
2026 obban
2027 obbet
2028 obbiz
2029 obbod
2030 obbum
2031 obbyc
2032 obcap
2033 obces
2034 obciv
2035 obcof
2036 obcuc
2037 obcyx
2038 obdez
2039 obdip
2040 obdog
2041 obdux
2042 obeat
2043 obedy
2044 obefo
2045 obelv
2046 obemu
2047 oberg
2048 obfas
2049 obfix
2050 obfor
2051 obfug
2052 obgav
2053 obgim
2054 obgop
2055 obgur
2056 obgyn
2057 obhab
2058 obhek
2059 obhig
2060 obhon
2061 obhup
2062 obiny
2063 obipu
2064 obiwa
2065 objah
2066 objer
2067 objis
2068 obkaz
2069 obkep
2070 obkin
2071 obkok
2072 obkus
2073 obkym
2074 oblal
2075 oblex
2076 oblib
2077 obloh
2078 oblut
2079 oblyr
2080 obmac
2081 obmeg
2082 obmif
2083 obmom
2084 obnew
2085 obnik
2086 obnol
2087 obobo
2088 obofy
2089 obome
2090 obonc
2091 obopa
2092 obots
2093 obpaf
2094 obped
2095 obpil
2096 obpoc
2097 obpuz
2098 obrag
2099 obreb
2100 obrid
2101 obrox
2102 obrun
2103 obryf
2104 obsaw
2105 obsef
2106 obsit
2107 obsoz
2108 obsuv
2109 obsys
2110 obten
2111 obtha
2112 obtry
2113 obtuk
2114 obuba
2115 obufe
2116 obuku
2117 obuni
2118 obuzo
2119 obvak
2120 obvic
2121 obvul
2122 obweh
2123 obyle
2124 obyub
2125 obzam
2126 obzel
2127 obzoy
2128 ocabu
2129 ocaca
2130 ocads
2131 ocafi
2132 ocago
2133 ocalm
2134 ocane
2135 ocaty
2136 ocawk
2137 ocbli
2138 occav
2139 occit
2140 occop
2141 occyg
2142 ocdax
2143 ocder
2144 ocdiz
2145 ocdov
2146 ocdup
2147 ocemi
2148 ocepa
2149 ocero
2150 ocest
2151 ocewe
2152 ocfaz
2153 ocfep
2154 ocfin
2155 ocfly
2156 ocfok
2157 ocfus
2158 ocgab
2159 ocgev
2160 ocgis
2161 ocgon
2162 ocgyr
2163 ochap
2164 ochof
2165 ochuc
2166 ochyd
2167 ocibs
2168 ociod
2169 ocirl
2170 ocjac
2171 ocjeg
2172 ocjuw
2173 ockad
2174 ockec
2175 ockir
2176 ockol
2177 ockub
2178 oclaf
2179 ocled
2180 oclip
2181 ocloc
2182 ocluz
2183 oclyn
2184 ocmag
2185 ocmeb
2186 ocmil
2187 ocmox
2188 ocmun
2189 ocmyc
2190 ocnah
2191 ocnef
2192 ocnib
2193 ocnoz
2194 ocnyl
2195 ococi
2196 ocoky
2197 ocord
2198 ocown
2199 ocoxo
2200 ocpak
2201 ocpem
2202 ocpid
2203 ocplo
2204 ocpry
2205 ocpuf
2206 ocral
2207 ocren
2208 ocrij
2209 ocros
2210 ocrut
2211 ocsam
2212 ocsel
2213 ocsho
2214 ocsif
2215 ocsud
2216 ocsyk
2217 octat
2218 octez
2219 octig
2220 octoy
2221 octru
2222 ocufo
2223 ocuma
2224 ocund
2225 ocuse
2226 ocuti
2227 ocvan
2228 ocvik
2229 ocvum
2230 ocwar
2231 ocwes
2232 ocwot
2233 oczey
2234 odadu
2235 odafa
2236 odale
2237 odani
2238 odaph
2239 odarx
2240 odato
2241 odbak
2242 odbem
2243 odbid
2244 odbob
2245 odbuf
2246 odbyx
2247 odcaw
2248 odcoy
2249 odcut
2250 oddac
2251 oddeg
2252 oddul
2253 oddyp
2254 odefi
2255 odega
2256 odeke
2257 odelm
2258 odepu
2259 odery
2260 odeth
2261 odfam
2262 odfel
2263 odfig
2264 odfot
2265 odfud
2266 odgan
2267 odgew
2268 odgic
2269 odgod
2270 odgum
2271 odhay
2272 odhet
2273 odhor
2274 odhym
2275 odici
2276 odigy
2277 odiku
2278 odilo
2279 odima
2280 odise
2281 odjap
2282 odjes
2283 odjil
2284 odjof
2285 odjux
2286 odkar
2287 odkib
2288 odkog
2289 odlas
2290 odlek
2291 odlim
2292 odlow
2293 odlug
2294 odlyb
2295 odmaj
2296 odmeh
2297 odmik
2298 odmop
2299 odmur
2300 odmyd
2301 odnax
2302 odner
2303 odnif
2304 odnov
2305 odnup
2306 odnyg
2307 odoch
2308 odode
2309 odoja
2310 odomi
2311 odops
2312 odorf
2313 odotu
2314 odpaz
2315 odpep
2316 odpit
2317 odpok
2318 odrab
2319 odrev
2320 odriz
2321 odron
2322 odruk
2323 odsad
2324 odsec
2325 odsol
2326 odspy
2327 odsub
2328 odted
2329 odtip
2330 odtox
2331 odtuc
2332 odtyt
2333 odubo
2334 oduka
2335 odulu
2336 odune
2337 odvat
2338 odvix
2339 odvoc
2340 odwal
2341 odwiv
2342 odxis
2343 odyom
2344 ofagy
2345 ofala
2346 ofano
2347 ofare
2348 ofasi
2349 ofatu
2350 ofawl
2351 ofbad
2352 ofbec
2353 ofbik
2354 ofbol
2355 ofbub
2356 ofcaf
2357 ofced
2358 ofchi
2359 ofcoc
2360 ofcuz
2361 ofcyk
2362 ofdag
2363 ofdeb
2364 ofdid
2365 ofdox
2366 ofdun
2367 ofeky
2368 ofeld
2369 ofepi
2370 oferu
2371 ofete
2372 offat
2373 offeh
2374 offoz
2375 ofgak
2376 ofgem
2377 ofgly
2378 ofgob
2379 ofguf
2380 ofhil
2381 ofhow
2382 ofibu
2383 ofify
2384 ofige
2385 ofind
2386 ofirk
2387 ofisl
2388 ofivo
2389 ofixa
2390 ofjal
2391 ofjen
2392 ofjos
2393 ofjut
2394 ofkam
2395 ofkel
2396 ofkif
2397 ofkro
2398 ofkud
2399 oflan
2400 oflet
2401 oflis
2402 oflod
2403 oflum
2404 ofmap
2405 ofmes
2406 ofmiz
2407 ofmof
2408 ofmuc
2409 ofmyt
2410 ofnac
2411 ofneg
2412 ofniv
2413 ofnom
2414 ofnys
2415 ofoce
2416 ofodo
2417 ofoli
2418 ofoms
2419 ofonz
2420 ofory
2421 ofoth
2422 ofowa
2423 ofoxu
2424 ofpax
2425 ofper
2426 ofpic
2427 ofpov
2428 ofpup
2429 ofrav
2430 ofrex
2431 ofrip
2432 ofrot
2433 ofrul
2434 ofrym
2435 ofsar
2436 ofsez
2437 ofsin
2438 ofsog
2439 ofsux
2440 oftaz
2441 oftyr
2442 ofudi
2443 ofugu
2444 ofuko
2445 ofule
2446 ofuna
2447 ofvas
2448 ofvib
2449 ofvor
2450 ofwah
2451 ofwey
2452 ofyra
2453 ofzab
2454 ofzef
2455 ofzit
2456 ogaby
2457 ogadz
2458 ogaho
2459 ogalc
2460 ogana
2461 ogari
2462 ogasm
2463 ogaxu
2464 ogbag
2465 ogbef
2466 ogbib
2467 ogbox
2468 ogbun
2469 ogbyl
2470 ogcay
2471 ogcet
2472 ogcoz
2473 ogdak
2474 ogdem
2475 ogdig
2476 ogdob
2477 ogduf
2478 ogdyn
2479 ogebo
2480 ogede
2481 ogefy
2482 ogela
2483 ogern
2484 ogeti
2485 ogfen
2486 ogfic
2487 ogfoy
2488 ogfut
2489 oggam
2490 oggel
2491 oggip
2492 oggot
2493 oggru
2494 oghaf
2495 ogher
2496 oghis
2497 oghod
2498 oghuz
2499 ogidi

2500 ogifu	**2550** ogtal	**2600** ohtle	**2650** ojido	**2700** ojtis	**2750** okguz	**2800** oktak	**2850** olfip	**2900** olrow	**2950** omele
2501 ogilk	2551 ogteg	2601 ohtro	2651 ojine	2701 ojtuf	2751 okhax	2801 oktet	2851 olflu	2901 olryn	2951 omeph
2502 ogisa	2552 ogtin	2602 ohuel	2652 ojisi	2702 ojubs	2752 okick	2802 oktij	2852 olfon	2902 olsew	2952 omerd
2503 ogite	2553 ogtos	2603 ohuha	2653 ojiza	2703 ojuck	2753 okidy	2803 oktoc	2853 olgac	2903 olsil	2953 ometo
2504 ogivy	2554 ogtum	2604 ohury	2654 ojkas	2704 ojuly	2754 okifi	2804 oktun	2854 olgeg	2904 olsod	2954 omewi
2505 ogjan	2555 ogtyx	2605 ohvop	2655 ojkek	2705 ojuno	2755 okips	2805 oktyd	2855 olgin	2905 olsum	2955 omfac
2506 ogjeb	2556 ogucu	2606 ohwas	2656 ojkid	2706 ojura	2756 okire	2806 okugi	2856 olgom	2906 oltas	2956 omfeg
2507 ogjik	2557 oguki	2607 ohwep	2657 ojkor	2707 ojuzi	2757 okita	2807 okunu	2857 olgyl	2907 oltik	2957 omfib
2508 ogkap	2558 ogung	2608 ohwib	2658 ojkug	2708 ojvag	2758 okjul	2808 okupo	2858 olhad	2908 oltol	2958 omfom
2509 ogkes	2559 ogupa	2609 ohzun	2659 ojkyp	2709 ojvil	2759 okker	2809 okurb	2859 olhif	2909 oltut	2959 omfru
2510 ogkim	**2560** oguso	**2610** ojaco	**2660** ojlaw	**2710** ojvox	**2760** okkig	**2810** okute	**2860** olhot	**2910** olufa	**2960** omgad
2511 ogkuc	2561 ogvad	2611 ojada	2661 ojleh	2711 ojvun	2761 okkup	2811 okveh	2861 olhug	2911 olunc	2961 omgec
2512 oglar	2562 ogvec	2612 ojame	2662 ojlit	2712 ojwir	2762 okkyf	2812 okvip	2862 olhyr	2912 olurg	2962 omgif
2513 oglez	2563 ogvir	2613 ojans	2663 ojlur	2713 ojwob	2763 oklem	2813 okvow	2863 olicy	2913 olust	2963 omgol
2514 oglid	2564 ogvol	2614 ojarg	2664 ojlyd	2714 okafy	2764 oklin	2814 okvus	2864 oligi	2914 oluxo	2964 omgub
2515 oglog	2565 ogwac	2615 ojavu	2665 ojmax	2715 okals	2765 oklob	2815 okvyt	2865 olinz	2915 oluze	2965 omhav
2516 oglux	2566 ogwed	2616 ojbal	2666 ojmer	2716 okamu	2766 okluf	2816 okwav	2866 oliro	2916 olvap	2966 omhit
2517 oglym	2567 ogyls	2617 ojben	2667 ojmib	2717 okank	2767 oklyx	2817 okwec	2867 olism	2917 olves	2967 omhoy
2518 ogmas	2568 ogyoc	2618 ojbix	2668 ojmov	2718 okapi	2768 okmal	2818 okwon	2868 oljim	2918 olviv	2968 omhul
2519 ogmek	2569 ogzat	2619 ojbos	2669 ojmut	2719 okart	2769 okmen	2819 okxol	2869 oljoy	2919 olwit	2969 omibe
2520 ogmit	**2570** ogzom	**2620** ojcab	**2670** ojmyn	**2720** okavo	**2770** okmos	**2820** okyba	**2870** oljub	**2920** olwor	**2970** omica
2521 ogmor	2571 ohaig	2621 ojcli	2671 ojnaz	2721 okawa	2771 okmyr	2821 okzid	2871 olkaf	2921 olxel	2971 omifo
2522 ogmug	2572 ohalt	2622 ojcon	2672 ojnep	2722 okaxe	2772 oknam	2822 okzle	2872 olked	2922 olyar	2972 omigu
2523 ognav	2573 ohaon	2623 ojcre	2673 ojnim	2723 okbaj	2773 oknel	2823 okzot	2873 olkli	2923 olyex	2973 omimp
2524 ogneh	2574 ohaum	2624 ojdam	2674 ojnus	2724 okbil	2774 oknit	2824 olact	2874 olkoc	2924 olyog	2974 omjaf
2525 ognop	2575 ohbra	2625 ojdel	2675 ojofo	2725 okboz	2775 oknud	2825 olady	2875 ollag	2925 olzek	2975 omjed
2526 ognur	2576 ohcyt	2626 ojdiv	2676 ojoke	2726 okbys	2776 oknyc	2826 olama	2876 olleb	2926 olzir	2976 omjoc
2527 ogobe	2577 ohdow	2627 ojdro	2677 ojold	2727 okcan	2777 okocu	2827 olapo	2877 olliz	2927 omaba	2977 omkag
2528 ogock	2578 ohdre	2628 ojdud	2678 ojont	2728 okcey	2778 okoko	2828 olaux	2878 ollox	2928 omach	2978 omkeb
2529 ogolo	2579 ohead	2629 ojdyt	2679 ojopu	2729 okcod	2779 okoma	2829 olbav	2879 ollun	2929 omaim	2979 omkil
2530 ogomu	**2580** ohelo	**2630** ojebe	**2680** ojorb	**2730** okcum	**2780** okope	**2830** olbij	**2880** ollyt	**2930** omaju	**2980** omkun
2531 ogoni	2581 oherb	2631 ojend	2681 ojovi	2731 okdaw	2781 okosy	2831 olboh	2881 olmah	2931 omaly	2981 omkyd
2532 ogopy	2582 ohfuc	2632 ojerf	2682 ojoxa	2732 okdex	2782 okoti	2832 olbur	2882 olnem	2932 omant	2982 omlay
2533 ogota	2583 ohgah	2633 ojewo	2683 ojpac	2733 okdis	2783 okpas	2833 olcax	2883 olnid	2933 omase	2983 omlef
2534 ogpab	2584 ohkom	2634 ojfan	2684 ojpeg	2734 okdly	2784 okpek	2834 olcer	2884 olnob	2934 omazo	2984 omloz
2535 ogpev	2585 ohmym	2635 ojfew	2685 ojpin	2735 okdur	2785 okpix	2835 olcib	2885 olnuf	2935 ombax	2985 omluv
2536 ogpif	2586 ohnat	2636 ojfod	2686 ojplu	2736 okejo	2786 okpor	2836 olcov	2886 oloak	2936 omblo	2986 ommak
2537 ogpon	2587 ohofe	2637 ojfri	2687 ojpom	2737 okelf	2787 okpug	2837 olcup	2887 olofu	2937 ombup	2987 ommem
2538 ogpuk	2588 ohohn	2638 ojfum	2688 ojrad	2738 okene	2788 okrap	2838 oldaz	2888 ololy	2938 ombyn	2988 ommip
2539 ograx	2589 ohoty	2639 ojgap	2689 ojrec	2739 okeva	2789 okres	2839 oldep	2889 olone	2939 omcaz	2989 ommob
2540 ogrhe	**2590** ohpri	**2640** ojges	**2690** ojrol	**2740** okexu	**2790** okriv	**2840** oldix	**2890** olort	**2940** omcep	**2990** ommuf
2541 ogrov	2591 ohret	2641 ojgiz	2691 ojrub	2741 okfar	2791 okrof	2841 oldok	2891 olosi	2941 omcro	2991 ommyg
2542 ogrup	2592 ohrin	2642 ojgof	2692 ojsaf	2742 okfez	2792 okruc	2842 oledi	2892 olowo	2942 omdab	2992 omnal
2543 ogryd	2593 ohroy	2643 ojguc	2693 ojsed	2743 okfik	2793 okryb	2843 oleng	2893 olpal	2943 omdev	2993 omnek
2544 ogsaz	2594 ohsca	2644 ojhak	2694 ojshy	2744 okfog	2794 oksac	2844 olerl	2894 olpen	2944 omdic	2994 omnix
2545 ogsep	2595 ohshe	2645 ojhey	2695 ojsig	2745 okfux	2795 okseg	2845 olesy	2895 olpig	2945 omdon	2995 omnyo
2546 ogsmi	2596 ohsny	2646 ojhip	2696 ojsoc	2746 okgaf	2796 oksir	2846 olewa	2896 olpos	2946 omdry	2996 omobs
2547 ogsow	2597 ohsok	2647 ojhoz	2697 ojsuz	2747 okged	2797 oksom	2847 olexe	2897 olrat	2947 omduk	2997 omoda
2548 ogsty	2598 ohsut	2648 ojhux	2698 ojtav	2748 okgno	2798 oksuk	2848 olfab	2898 olrez	2948 omebu	2998 omofi
2549 ogsus	2599 ohtay	2649 ojich	2699 ojtem	2749 okgry	2799 oksyn	2849 olfev	2899 olrhi	2949 omefa	2999 omojo

omoku omowl ompiv ompud

[redacted]

SECTION B

3000 omran	**3050** ondil	**3100** onnot	**3150** opcob	**3200** opoki	**3250** oqcip	**3300** orape	**3350** oroga	**3400** osews	**3450** oswaw
3001 omrew	3051 ondoc	3101 onobu	3151 opcra	3201 opolt	3251 oqcly	3301 orara	3351 orolu	3401 osfro	3451 oszúg
3002 omrik	3052 onduz	3102 onogy	3152 opcyc	3202 opons	3252 oqcru	3302 orbaf	3352 oropi	3402 osful	3452 otaep
3003 omrod	3053 ondyx	3103 onols	3153 opdap	3203 opowe	3253 oqdah	3303 orboc	3353 orovo	3403 osgez	3453 otand
3004 omrum	3054 onech	3104 onond	3154 opdes	3204 oppan	3254 oqdit	3304 orbry	3354 oroxe	3404 oshov	3454 otaus
3005 omryl	3055 onedo	3105 onorc	3155 opdof	3205 oppet	3255 oqdye	3305 orbuz	3355 orpav	3405 oshyg	3455 otbat
3006 omsap	3056 onege	3106 onoze	3156 opduc	3206 opphy	3256 oqeca	3306 orday	3356 orpex	3406 osiba	3456 otbeb
3007 omses	3057 oneki	3107 onpar	3157 opeha	3207 oppib	3257 oqegy	3307 ordef	3357 orpli	3407 osife	3457 otbir
3008 omsiz	3058 onent	3108 onpez	3158 openy	3208 oppod	3258 oqels	3308 ordoz	3358 orpop	3408 osimi	3458 otbow
3009 omsof	3059 onesu	3109 onpik	3159 opepo	3209 oppum	3259 oqeum	3309 orduv	3359 orpyk	3409 osiru	3459 otbul
3010 omsuc	**3060** onfag	**3110** onpog	**3160** opets	**3210** oprac	**3260** oqezo	**3310** orect	**3360** orraz	**3410** osjas	**3460** otcal
3011 omsyr	3061 onfeb	3111 onpyt	3161 opexi	3211 opreg	3261 oqfak	3311 oregu	3361 orrep	3411 osjor	3461 otcef
3012 omtaw	3062 onfir	3112 onrib	3162 opeze	3212 oprom	3262 oqfle	3312 oreko	3362 orrok	3412 oskav	3462 otcim
3013 omtex	3063 onfox	3113 onsav	3163 opfay	3213 opsat	3263 oqhas	3313 orems	3363 orryc	3413 oskew	3463 otcot
3014 omthi	3064 onfun	3114 onsim	3164 opfex	3214 opsoy	3264 oqhed	3314 orepy	3364 orsab	3414 oskiz	3464 otdad
3015 omtor	3065 ongat	3115 onsly	3165 opfil	3215 opsul	3265 oqhuw	3315 orexa	3365 orsev	3415 oskry	3465 otdec
3016 omtuy	3066 ongef	3116 onsop	3166 opfoh	3216 optax	3266 oqila	3316 orezi	3366 orson	3416 osler	3466 otdol
3017 omuga	3067 ongid	3117 onsur	3167 opfuz	3217 optid	3267 oqipe	3317 orfah	3367 orsup	3417 oslup	3467 otdub
3018 omumi	3068 ongoz	3118 ontab	3168 opgas	3218 optup	3268 oqito	3318 orfem	3368 orsym	3418 osmaz	3468 otdym
3019 omurn	3069 ongul	3119 onthy	3169 opgek	3219 opuby	3269 oqkoy	3319 orfyl	3369 ortec	3419 osmin	3469 oteda
3020 omutu	**3070** ongyb	**3120** ontit	**3170** opgit	**3220** opufu	**3270** oqkra	**3320** orgen	**3370** ortir	**3420** osmok	**3470** otemb
3021 omvar	3071 onhok	3121 ontow	3171 opgor	3221 opujo	3271 oqkuv	3321 orgiv	3371 ortub	3421 osnev	3471 oteni
3022 omvez	3072 onhyc	3122 ontre	3172 opgug	3222 opulk	3272 oqlew	3322 orgos	3372 ortyp	3422 osnig	3472 oteps
3023 omvin	3073 onibi	3123 onube	3173 ophec	3223 opvaz	3273 oqmet	3323 orgut	3373 orubi	3423 osnuk	3473 oterz
3024 omvog	3074 onicu	3124 onuja	3174 ophiv	3224 opvey	3274 oqmow	3324 orheh	3374 orucy	3424 osnyp	3474 otetu
3025 omwat	3075 onida	3125 onult	3175 opids	3225 opvim	3275 oqmuy	3325 orhuy	3375 orulo	3425 osody	3475 otexo
3026 omwid	3076 oniko	3126 onumy	3176 opiga	3226 opvok	3276 oqnaw	3326 oridu	3376 orups	3426 osola	3476 oteye
3027 omyah	3077 oning	3127 onunz	3177 opinu	3227 opvuf	3277 oqnen	3327 orily	3377 oruxu	3427 osomo	3477 otfed
3028 omyen	3078 onith	3128 onvov	3178 opipy	3228 opwag	3278 oqnuc	3328 orino	3378 orvac	3428 osong	3478 otfis
3029 omyut	3079 onjak	3129 onvup	3179 opiri	3229 opweb	3279 oqoks	3329 oriva	3379 orveg	3429 osose	3479 otgaw
3030 omzop	**3080** onjem	**3130** onwaz	**3180** opixo	**3230** opwin	**3280** oqona	**3330** orize	**3380** orvom	**3430** ospec	**3480** otget
3031 onabo	3081 onjob	3131 onwet	3181 opjaw	3231 opwow	3281 oqpat	3331 orjel	3381 orwak	3431 ospij	3481 otgig
3032 onacs	3082 onjuf	3132 onyew	3182 opkef	3232 opwry	3282 oqpoh	3332 orjig	3382 orwic	3432 ospol	3482 otgro
3033 onald	3083 onkal	3133 onyla	3183 opkoz	3233 opxal	3283 oqpul	3333 orjus	3383 orzax	3433 ospub	3483 otguy
3034 onapy	3084 onken	3134 onyus	3184 opkru	3234 opyaf	3284 oqrer	3334 orkey	3384 orzer	3434 osraf	3484 othan
3035 onarm	3085 onkyk	3135 onzah	3185 oplah	3235 opyed	3285 oqrha	3335 orkip	3385 osaci	3435 osril	3485 othoc
3036 onate	3086 onlam	3136 onzin	3186 oplen	3236 opyop	3286 oqrim	3336 orkod	3386 osafu	3436 osruz	3486 otiha
3037 onauk	3087 onlel	3137 opacu	3187 oplos	3237 opzad	3287 oqsey	3337 orkum	3387 osbiv	3437 osseb	3487 otims
3038 onawn	3088 onlix	3138 opams	3188 opluw	3238 opzig	3288 oqshi	3338 orlap	3388 oscac	3438 ossyd	3488 otirp
3039 onaxi	3089 onlon	3139 oparo	3189 opmam	3239 opzol	3289 oqsma	3339 orles	3389 osceg	3439 ostah	3489 otitz
3040 onaza	**3090** onlud	**3140** opata	**3190** opmel	**3240** opzub	**3290** oqsot	**3340** orluc	**3390** oscom	**3440** ostob	**3490** otjod
3041 onbif	3091 onman	3141 opave	3191 opmir	3241 oqako	3291 oqthe	3341 ormar	3391 oscys	3441 ostyn	3491 otjum
3042 onblu	3092 onmex	3142 opaxy	3192 opmot	3242 oqats	3292 oqtib	3342 ormez	3392 osdan	3442 osuco	3492 otkay
3043 onbro	3093 onmig	3143 opazi	3193 opmud	3243 oqaug	3293 oqtok	3343 ormid	3393 osdey	3443 osuky	3493 otkex
3044 oncec	3094 onmoh	3144 opbab	3194 opnar	3244 oqaxa	3294 oqtud	3344 ormog	3394 osdif	3444 osume	3494 otklo
3045 oncol	3095 onmum	3145 opbev	3195 opnog	3245 oqaze	3295 oqwap	3345 ormux	3395 osdod	3445 osuph	3495 otkut
3046 oncub	3096 onmyl	3146 opbis	3196 opnux	3246 oqbay	3296 oqxav	3346 ormys	3396 osdut	3446 osuzu	3496 otlab
3047 oncyn	3097 onnap	3147 opbon	3197 opnyk	3247 oqbom	3297 orahy	3347 ornas	3397 oseja	3447 osvay	3497 otley
3048 ondaf	3098 onnes	3148 opbuk	3198 opoco	3248 oqbri	3298 oraix	3348 ornug	3398 oseli	3448 osvef	3498 otlop
3049 onded	3099 onnic	3149 opciz	3199 opohm	3249 oqcho	3299 orang	3349 ornyx	3399 oserx	3449 osvoz	3499 otluk

3500 otmew	**3550** ovcyp	**3600** ovney	**3650** owaka	**3700** owmyx	**3750** owzon	**3800** oxoha	**3850** oygli	**3900** oyvon	**3950** ozkab
3501 otmoy	3551 ovdal	3601 ovnod	3651 owalo	3701 ownir	3751 owzuf	3801 oxove	3851 oygoc	3901 oywaf	3951 ozkev
3502 otnin	3552 ovden	3602 ovnum	3652 owamp	3702 ownul	3752 oxanz	3802 oxoys	3852 oygre	3902 oywip	3952 ozkis
3503 otnos	3553 ovdik	3603 ovogi	3653 owanu	3703 owony	3753 oxapa	3803 oxpir	3853 oygud	3903 oywog	3953 ozklu
3504 otnyt	3554 ovdos	3604 ovomy	3654 oward	3704 owoph	3754 oxask	3804 oxpoz	3854 oyhez	3904 oywye	3954 ozkon
3505 otoaf	3555 ovduy	3605 ovono	3655 owasy	3705 owore	3755 oxath	3805 oxrak	3855 oyimy	3905 oyxar	3955 ozkyr
3506 otoen	3556 ovecy	3606 ovork	3656 owaxt	3706 owoso	3756 oxauc	3806 oxrem	3856 oyini	3906 oyxen	3956 ozlac
3507 otoid	3557 oveka	3607 ovosa	3657 owbah	3707 owova	3757 oxaxo	3807 oxrob	3857 oyizo	3907 oyxim	3957 ozleg
3508 otora	3558 ovelb	3608 ovovu	3658 owbey	3708 owozi	3758 oxbez	3808 oxruf	3858 oykaw	3908 ozahs	3958 ozlif
3509 otoux	3559 ovenu	3609 ovpap	3659 owbot	3709 owpew	3759 oxbim	3809 oxscu	3859 oykem	3909 ozalb	3959 ozlom
3510 otows	**3560** ovesi	**3610** ovpes	**3660** owcaj	**3710** owpha	**3760** oxbly	**3810** oxski	**3860** oykot	**3910** ozamt	**3960** ozlyp
3511 otozo	3561 ovfad	3611 ovpiz	3661 owcid	3711 owpis	3761 oxcin	3811 oxsos	3861 oykuk	3911 ozaud	3961 ozmad
3512 otpip	3562 ovfec	3612 ovpro	3662 owcos	3712 owple	3762 oxcor	3812 oxtiz	3862 oylav	3912 ozawe	3962 ozmec
3513 otpof	3563 ovflo	3613 ovpuc	3663 owdat	3713 owpoy	3763 oxcug	3813 oxtyl	3863 oylep	3913 ozaya	3963 ozmix
3514 otpre	3564 ovfry	3614 ovpyd	3664 owdru	3714 owram	3764 oxcyb	3814 oxuca	3864 oylir	3914 ozbap	3964 ozmol
3515 otrah	3565 ovfub	3615 ovrar	3665 owean	3715 owrel	3765 oxdav	3815 oxunt	3865 oylok	3915 ozbes	3965 oznaf
3516 otrog	3566 ovgag	3616 ovrit	3666 owefu	3716 owroj	3766 oxdej	3816 oxvam	3866 oylyc	3916 ozbig	3966 ozned
3517 otrys	3567 ovgir	3617 ovryx	3667 owely	3717 owruw	3767 oxdop	3817 oxvel	3867 oymat	3917 ozbof	3967 oznoc
3518 otsak	3568 ovgle	3618 ovsas	3668 owerk	3718 owsax	3768 oxegi	3818 oxvij	3868 oymey	3918 ozbuc	3968 oznuz
3519 otshu	3569 ovgun	3619 ovsek	3669 owevo	3719 owske	3769 oxema	3819 oxvot	3869 oymod	3919 ozbyb	3969 ozoca
3520 otsme	**3570** ovhah	**3620** ovsor	**3670** owfal	**3720** owsli	**3770** oxeno	**3820** oxyat	**3870** oymup	**3920** ozcar	**3970** ozodi
3521 ottek	3571 ovheg	3621 ovspi	3671 owfet	3721 owsov	3771 oxepe	3821 oxzeh	3871 oynak	3921 ozchy	3971 ozonu
3522 ottho	3572 ovhol	3622 ovsug	3672 owfif	3722 owspu	3772 oxert	3822 oxzof	3872 oynec	3922 ozcic	3972 ozost
3523 ottuz	3573 ovhur	3623 ovsyl	3673 owfoc	3723 owsyc	3773 oxezu	3823 oyaam	3873 oyniz	3923 ozcog	3973 ozoxy
3524 otusi	3574 oviby	3624 ovtac	3674 owgeh	3724 owtaf	3774 oxgep	3824 oyabs	3874 oynof	3924 ozcux	3974 ozpag
3525 otvur	3575 ovifa	3625 ovtof	3675 owgow	3725 owtes	3775 oxgil	3825 oyajo	3875 oynut	3925 ozdek	3975 ozpeb
3526 otwam	3576 ovils	3626 ovtye	3676 owgra	3726 owtic	3776 oxgus	3826 oyake	3876 oyoby	3926 ozdor	3976 ozpim
3527 otwer	3577 ovitu	3627 ovugo	3677 owhag	3727 owtog	3777 oxhaz	3827 oyalu	3877 oyocs	3927 ozdug	3977 ozpow
3528 otwiz	3578 ovive	3628 ovuli	3678 owhem	3728 owtym	3778 oxhyp	3828 oyarc	3878 oyogn	3928 ozdyc	3978 ozpun
3529 otxem	3579 ovizi	3629 ovumu	3679 owhop	3729 owuch	3779 oxili	3829 oyavi	3879 oyoka	3929 ozelz	3979 ozraw
3530 otyap	**3580** ovjay	**3630** ovush	**3680** owhus	**3730** owudu	**3780** oximu	**3830** oybex	**3880** oyole	**3930** ozemy	**3980** ozref
3531 otyel	3581 ovjef	3631 ovuta	3681 owhyb	3731 owunk	3781 oxina	3831 oycig	3881 oyoro	3931 ozept	3981 ozroz
3532 otzez	3582 ovjid	3632 ovvav	3682 owiar	3732 owuri	3782 oxipo	3832 oycul	3882 oyotz	3932 ozevu	3982 ozsah
3533 otzor	3583 ovjot	3633 ovvet	3683 owied	3733 owuto	3783 oxist	3833 oycym	3883 oypaj	3933 ozfav	3983 ozsem
3534 otzuc	3584 ovjup	3634 ovvif	3684 owiky	3734 owuxa	3784 oxjab	3834 oydet	3884 oypho	3934 ozfey	3984 ozsip
3535 ovade	3585 ovkak	3635 ovvoy	3685 owins	3735 owvex	3785 oxjif	3835 oydom	3885 oypuy	3935 ozfiz	3985 ozsob
3536 ovaku	3586 ovkeh	3636 ovwax	3686 owipa	3736 owvig	3786 oxjon	3836 oydri	3886 oyray	3936 ozfop	3986 ozsuf
3537 ovany	3587 ovkob	3637 ovwig	3687 owjaz	3737 owvyn	3787 oxkac	3837 oydyl	3887 oyroh	3937 ozfur	3987 oztam
3538 ovapt	3588 ovkuf	3638 ovwom	3688 owjec	3738 owwab	3788 oxkyn	3838 oyebi	3888 oysko	3938 ozgax	3988 oztep
3539 ovars	3589 ovkys	3639 ovwyn	3689 owjin	3739 owwen	3789 oxlec	3839 oyeft	3889 oysla	3939 ozger	3989 oztiv
3540 ovbaz	**3590** ovlat	**3640** ovyca	**3690** owjol	**3740** owwil	**3790** oxliv	**3840** oyelk	**3890** oystu	**3940** ozgib	**3990** oztot
3541 ovbin	3591 ovlov	3641 ovyem	3691 owjud	3741 owwox	3791 oxlol	3841 oyeme	3891 oytan	3941 ozgov	3991 oztul
3542 ovbok	3592 ovlul	3642 ovyux	3692 owkik	3742 owxas	3792 oxlub	3842 oyena	3892 oytop	3942 ozgup	3992 ozudy
3543 ovbre	3593 ovlyg	3643 ovzaf	3693 owkof	3743 owxer	3793 oxmaf	3843 oyerp	3893 oytux	3943 ozhaj	3993 ozval
3544 ovbus	3594 ovmab	3644 ovzed	3694 owkux	3744 owyez	3794 oxmed	3844 oyesh	3894 oyual	3944 ozhir	3994 ozven
3545 ovbyt	3595 ovmev	3645 ovzib	3695 owlak	3745 owyok	3795 oxmoc	3845 oyeuc	3895 oyubu	3945 ozhox	3995 ozvit
3546 ovcam	3596 ovmim	3646 ovzly	3696 owlev	3746 owyuc	3796 oxmuz	3846 oyfol	3896 oyude	3946 ozhuk	3996 ozvos
3547 ovcel	3597 ovmon	3647 ovzoc	3697 owlor	3747 owzac	3797 oxnox	3847 oyfra	3897 oyufi	3947 ozigo	3997 ozway
3548 ovcow	3598 ovmuk	3648 ovzuz	3698 owlys	3748 owzep	3798 oxodu	3848 oyfuf	3898 oyuor	3948 oziph	3998 ozwex
3549 ovcud	3599 ovnan	3649 owahi	3699 owmaw	3749 owzim	3799 oxogo	3849 oygaz	3899 oyvis	3949 ozizu	3999 ozxin

ozzan ozzik ozzum paafo

SECTION B

3000 omran	**3050** ondil	**3100** onnot	**3150** opcob	**3200** opoki	**3250** oqcip	**3300** orape	**3350** oroga	**3400** osews	**3450** oswaw
3001 omrew	3051 ondoc	3101 onobu	3151 opcra	3201 opolt	3251 oqcly	3301 orara	3351 orolu	3401 osfro	3451 oszúg
3002 omrik	3052 onduz	3102 onogy	3152 opcyc	3202 opons	3252 oqcru	3302 orbaf	3352 oropi	3402 osful	3452 otaep
3003 omrod	3053 ondyx	3103 onols	3153 opdap	3203 opowe	3253 oqdah	3303 orboc	3353 orovo	3403 osgez	3453 otand
3004 omrum	3054 onech	3104 onond	3154 opdes	3204 oppan	3254 oqdit	3304 orbry	3354 oroxe	3404 oshov	3454 otaus
3005 omryl	3055 onedo	3105 onorc	3155 opdof	3205 oppet	3255 oqdye	3305 orbuz	3355 orpav	3405 oshyg	3455 otbat
3006 omsap	3056 onege	3106 onoze	3156 opduc	3206 opphy	3256 oqeca	3306 orday	3356 orpex	3406 osiba	3456 otbeb
3007 omses	3057 oneki	3107 onpar	3157 opeha	3207 oppib	3257 oqegy	3307 ordef	3357 orpli	3407 osife	3457 otbir
3008 omsiz	3058 onent	3108 onpez	3158 openy	3208 oppod	3258 oqels	3308 ordoz	3358 orpop	3408 osimi	3458 otbow
3009 omsof	3059 onesu	3109 onpik	3159 opepo	3209 oppum	3259 oqeum	3309 orduv	3359 orpyk	3409 osiru	3459 otbul
3010 omsuc	**3060** onfag	**3110** onpog	**3160** opets	**3210** oprac	**3260** oqezo	**3310** orect	**3360** orraz	**3410** osjas	**3460** otcal
3011 omsyr	3061 onfeb	3111 onpyt	3161 opexi	3211 opreg	3261 oqfak	3311 oregu	3361 orrep	3411 osjor	3461 otcef
3012 omtaw	3062 onfir	3112 onrib	3162 opeze	3212 oprom	3262 oqfle	3312 oreko	3362 orrok	3412 oskav	3462 otcim
3013 omtex	3063 onfox	3113 onsav	3163 opfay	3213 opsat	3263 oqhas	3313 orems	3363 orryc	3413 oskew	3463 otcot
3014 omthi	3064 onfun	3114 onsim	3164 opfex	3214 opsoy	3264 oqhed	3314 orepy	3364 orsab	3414 oskiz	3464 otdad
3015 omtor	3065 ongat	3115 onsly	3165 opfil	3215 opsul	3265 oqhuw	3315 orexa	3365 orsev	3415 oskry	3465 otdec
3016 omtuy	3066 ongef	3116 onsop	3166 opfoh	3216 optax	3266 oqila	3316 orezi	3366 orson	3416 osler	3466 otdol
3017 omuga	3067 ongid	3117 onsur	3167 opfuz	3217 optid	3267 oqipe	3317 orfah	3367 orsup	3417 oslup	3467 otdub
3018 omumi	3068 ongoz	3118 ontab	3168 opgas	3218 optup	3268 oqito	3318 orfem	3368 orsym	3418 osmaz	3468 otdym
3019 omurn	3069 ongul	3119 onthy	3169 opgek	3219 opuby	3269 oqkoy	3319 orfyl	3369 ortec	3419 osmin	3469 oteda
3020 omutu	**3070** ongyb	**3120** ontit	**3170** opgit	**3220** opufu	**3270** oqkra	**3320** orgen	**3370** ortir	**3420** osmok	**3470** otemb
3021 omvar	3071 onhok	3121 ontow	3171 opgor	3221 opujo	3271 oqkuv	3321 orgiv	3371 ortub	3421 osnev	3471 oteni
3022 omvez	3072 onhyc	3122 ontre	3172 opgug	3222 opulk	3272 oqlew	3322 orgos	3372 ortyp	3422 osnig	3472 oteps
3023 omvin	3073 onibi	3123 onube	3173 ophec	3223 opvaz	3273 oqmet	3323 orgut	3373 orubi	3423 osnuk	3473 oterz
3024 omvog	3074 onicu	3124 onuja	3174 ophiv	3224 opvey	3274 oqmow	3324 orheh	3374 orucy	3424 osnyp	3474 otetu
3025 omwat	3075 onida	3125 onult	3175 opids	3225 opvim	3275 oqmuy	3325 orhuy	3375 orulo	3425 osody	3475 otexo
3026 omwid	3076 oniko	3126 onumy	3176 opiga	3226 opvok	3276 oqnaw	3326 oridu	3376 orups	3426 osola	3476 oteye
3027 omyah	3077 oning	3127 onunz	3177 opinu	3227 opvuf	3277 oqnen	3327 orily	3377 oruxu	3427 osomo	3477 otfed
3028 omyen	3078 onith	3128 onvov	3178 opipy	3228 opwag	3278 oqnuc	3328 orino	3378 orvac	3428 osong	3478 otfis
3029 omyut	3079 onjak	3129 onvup	3179 opiri	3229 opweb	3279 oqoks	3329 oriva	3379 orveg	3429 osose	3479 otgaw
3030 omzop	**3080** onjem	**3130** onwaz	**3180** opixo	**3230** opwin	**3280** oqona	**3330** orize	**3380** orvom	**3430** ospec	**3480** otget
3031 onabo	3081 onjob	3131 onwet	3181 opjaw	3231 opwow	3281 oqpat	3331 orjel	3381 orwak	3431 ospij	3481 otgig
3032 onacs	3082 onjuf	3132 onyew	3182 opkef	3232 opwry	3282 oqpoh	3332 orjig	3382 orwic	3432 ospol	3482 otgro
3033 onald	3083 onkal	3133 onyla	3183 opkoz	3233 opxal	3283 oqpul	3333 orjus	3383 orzax	3433 ospub	3483 otguy
3034 onapy	3084 onken	3134 onyus	3184 opkru	3234 opyaf	3284 oqrer	3334 orkey	3384 orzer	3434 osraf	3484 othan
3035 onarm	3085 onkyk	3135 onzah	3185 oplah	3235 opyed	3285 oqrha	3335 orkip	3385 osaci	3435 osril	3485 othoc
3036 onate	3086 onlam	3136 onzin	3186 oplen	3236 opyop	3286 oqrim	3336 orkod	3386 osafu	3436 osruz	3486 otiha
3037 onauk	3087 onlel	3137 opacu	3187 oplos	3237 opzad	3287 oqsey	3337 orkum	3387 osbiv	3437 osseb	3487 otims
3038 onawn	3088 onlix	3138 opams	3188 opluw	3238 opzig	3288 oqshi	3338 orlap	3388 oscac	3438 ossyd	3488 otirp
3039 onaxi	3089 onlon	3139 oparo	3189 opmam	3239 opzol	3289 oqsma	3339 orles	3389 osceg	3439 ostah	3489 otitz
3040 onaza	**3090** onlud	**3140** opata	**3190** opmel	**3240** opzub	**3290** oqsot	**3340** orluc	**3390** oscom	**3440** ostob	**3490** otjod
3041 onbif	3091 onman	3141 opave	3191 opmir	3241 oqako	3291 oqthe	3341 ormar	3391 oscys	3441 ostyn	3491 otjum
3042 onblu	3092 onmex	3142 opaxy	3192 opmot	3242 oqats	3292 oqtib	3342 ormez	3392 osdan	3442 osuco	3492 otkay
3043 onbro	3093 onmig	3143 opazi	3193 opmud	3243 oqaug	3293 oqtok	3343 ormid	3393 osdey	3443 osuky	3493 otkex
3044 oncec	3094 onmoh	3144 opbab	3194 opnar	3244 oqaxa	3294 oqtud	3344 ormog	3394 osdif	3444 osume	3494 otklo
3045 oncol	3095 onmum	3145 opbev	3195 opnog	3245 oqaze	3295 oqwap	3345 ormux	3395 osdod	3445 osuph	3495 otkut
3046 oncub	3096 onmyl	3146 opbis	3196 opnux	3246 oqbay	3296 oqxav	3346 ormys	3396 osdut	3446 osuzu	3496 otlab
3047 oncyn	3097 onnap	3147 opbon	3197 opnyk	3247 oqbom	3297 orahy	3347 ornas	3397 oseja	3447 osvay	3497 otley
3048 ondaf	3098 onnes	3148 opbuk	3198 opoco	3248 oqbri	3298 oraix	3348 ornug	3398 oseli	3448 osvef	3498 otlop
3049 onded	3099 onnic	3149 opciz	3199 opohm	3249 oqcho	3299 orang	3349 ornyx	3399 oserx	3449 osvoz	3499 otluk

3500 otmew	**3550** ovcyp	**3600** ovney	**3650** owaka	**3700** owmyx	**3750** owzon	**3800** oxoha	**3850** oygli	**3900** oyvon	**3950** ozkab
3501 otmoy	3551 ovdal	3601 ovnod	3651 owalo	3701 ownir	3751 owzuf	3801 oxove	3851 oygoc	3901 oywaf	3951 ozkev
3502 otnin	3552 ovden	3602 ovnum	3652 owamp	3702 ownul	3752 oxanz	3802 oxoys	3852 oygre	3902 oywip	3952 ozkis
3503 otnos	3553 ovdik	3603 ovogi	3653 owanu	3703 owony	3753 oxapa	3803 oxpir	3853 oygud	3903 oywog	3953 ozklu
3504 otnyt	3554 ovdos	3604 ovomy	3654 oward	3704 owoph	3754 oxask	3804 oxpoz	3854 oyhez	3904 oywye	3954 ozkon
3505 otoaf	3555 ovduy	3605 ovono	3655 owasy	3705 owore	3755 oxath	3805 oxrak	3855 oyimy	3905 oyxar	3955 ozkyr
3506 otoen	3556 ovecy	3606 ovork	3656 owaxt	3706 owoso	3756 oxauc	3806 oxrem	3856 oyini	3906 oyxen	3956 ozlac
3507 otoid	3557 oveka	3607 ovosa	3657 owbah	3707 owova	3757 oxaxo	3807 oxrob	3857 oyizo	3907 oyxim	3957 ozleg
3508 otora	3558 ovelb	3608 ovovu	3658 owbey	3708 owozi	3758 oxbez	3808 oxruf	3858 oykaw	3908 ozahs	3958 ozlif
3509 otoux	3559 ovenu	3609 ovpap	3659 owbot	3709 owpew	3759 oxbim	3809 oxscu	3859 oykem	3909 ozalb	3959 ozlom
3510 otows	**3560** ovesi	**3610** ovpes	**3660** owcaj	**3710** owpha	**3760** oxbly	**3810** oxski	**3860** oykot	**3910** ozamt	**3960** ozlyp
3511 otozo	3561 ovfad	3611 ovpiz	3661 owcid	3711 owpis	3761 oxcin	3811 oxsos	3861 oykuk	3911 ozaud	3961 ozmad
3512 otpip	3562 ovfec	3612 ovpro	3662 owcos	3712 owple	3762 oxcor	3812 oxtiz	3862 oylav	3912 ozawe	3962 ozmec
3513 otpof	3563 ovflo	3613 ovpuc	3663 owdat	3713 owpoy	3763 oxcug	3813 oxtyl	3863 oylep	3913 ozaya	3963 ozmix
3514 otpre	3564 ovfry	3614 ovpyd	3664 owdru	3714 owram	3764 oxcyb	3814 oxuca	3864 oylir	3914 ozbap	3964 ozmol
3515 otrah	3565 ovfub	3615 ovrar	3665 owean	3715 owrel	3765 oxdav	3815 oxunt	3865 oylok	3915 ozbes	3965 oznaf
3516 otrog	3566 ovgag	3616 ovrit	3666 owefu	3716 owroj	3766 oxdej	3816 oxvam	3866 oylyc	3916 ozbig	3966 ozned
3517 otrys	3567 ovgir	3617 ovryx	3667 owely	3717 owruw	3767 oxdop	3817 oxvel	3867 oymat	3917 ozbof	3967 oznoc
3518 otsak	3568 ovgle	3618 ovsas	3668 owerk	3718 owsax	3768 oxegi	3818 oxvij	3868 oymey	3918 ozbuc	3968 oznuz
3519 otshu	3569 ovgun	3619 ovsek	3669 owevo	3719 owske	3769 oxema	3819 oxvot	3869 oymod	3919 ozbyb	3969 ozoca
3520 otsme	**3570** ovhah	**3620** ovsor	**3670** owfal	**3720** owsli	**3770** oxeno	**3820** oxyat	**3870** oymup	**3920** ozcar	**3970** ozodi
3521 ottek	3571 ovheg	3621 ovspi	3671 owfet	3721 owsov	3771 oxepe	3821 oxzeh	3871 oynak	3921 ozchy	3971 ozonu
3522 ottho	3572 ovhol	3622 ovsug	3672 owfif	3722 owspu	3772 oxert	3822 oxzof	3872 oynec	3922 ozcic	3972 ozost
3523 ottuz	3573 ovhur	3623 ovsyl	3673 owfoc	3723 owsyc	3773 oxezu	3823 oyaam	3873 oyniz	3923 ozcog	3973 ozoxy
3524 otusi	3574 oviby	3624 ovtac	3674 owgeh	3724 owtaf	3774 oxgep	3824 oyabs	3874 oynof	3924 ozcux	3974 ozpag
3525 otvur	3575 ovifa	3625 ovtof	3675 owgow	3725 owtes	3775 oxgil	3825 oyajo	3875 oynut	3925 ozdek	3975 ozpeb
3526 otwam	3576 ovils	3626 ovtye	3676 owgra	3726 owtic	3776 oxgus	3826 oyake	3876 oyoby	3926 ozdor	3976 ozpim
3527 otwer	3577 ovitu	3627 ovugo	3677 owhag	3727 owtog	3777 oxhaz	3827 oyalu	3877 oyocs	3927 ozdug	3977 ozpow
3528 otwiz	3578 ovive	3628 ovuli	3678 owhem	3728 owtym	3778 oxhyp	3828 oyarc	3878 oyogn	3928 ozdyc	3978 ozpun
3529 otxem	3579 ovizi	3629 ovumu	3679 owhop	3729 owuch	3779 oxili	3829 oyavi	3879 oyoka	3929 ozelz	3979 ozraw
3530 otyap	**3580** ovjay	**3630** ovush	**3680** owhus	**3730** owudu	**3780** oximu	**3830** oybex	**3880** oyole	**3930** ozemy	**3980** ozref
3531 otyel	3581 ovjef	3631 ovuta	3681 owhyb	3731 owunk	3781 oxina	3831 oycig	3881 oyoro	3931 ozept	3981 ozroz
3532 otzez	3582 ovjid	3632 ovvav	3682 owiar	3732 owuri	3782 oxipo	3832 oycul	3882 oyotz	3932 ozevu	3982 ozsah
3533 otzor	3583 ovjot	3633 ovvet	3683 owied	3733 owuto	3783 oxist	3833 oycym	3883 oypaj	3933 ozfav	3983 ozsem
3534 otzuc	3584 ovjup	3634 ovvif	3684 owiky	3734 owuxa	3784 oxjab	3834 oydet	3884 oypho	3934 ozfey	3984 ozsip
3535 ovade	3585 ovkak	3635 ovvoy	3685 owins	3735 owvex	3785 oxjif	3835 oydom	3885 oypuy	3935 ozfiz	3985 ozsob
3536 ovaku	3586 ovkeh	3636 ovwax	3686 owipa	3736 owvig	3786 oxjon	3836 oydri	3886 oyray	3936 ozfop	3986 ozsuf
3537 ovany	3587 ovkob	3637 ovwig	3687 owjaz	3737 owvyn	3787 oxkac	3837 oydyl	3887 oyroh	3937 ozfur	3987 oztam
3538 ovapt	3588 ovkuf	3638 ovwom	3688 owjec	3738 owwab	3788 oxkyn	3838 oyebi	3888 oysko	3938 ozgax	3988 oztep
3539 ovars	3589 ovkys	3639 ovwyn	3689 owjin	3739 owwen	3789 oxlec	3839 oyeft	3889 oysla	3939 ozger	3989 oztiv
3540 ovbaz	**3590** ovlat	**3640** ovyca	**3690** owjol	**3740** owwil	**3790** oxliv	**3840** oyelk	**3890** oystu	**3940** ozgib	**3990** oztot
3541 ovbin	3591 ovlov	3641 ovyem	3691 owjud	3741 owwox	3791 oxlol	3841 oyeme	3891 oytan	3941 ozgov	3991 oztul
3542 ovbok	3592 ovlul	3642 ovyux	3692 owkik	3742 owxas	3792 oxlub	3842 oyena	3892 oytop	3942 ozgup	3992 ozudy
3543 ovbre	3593 ovlyg	3643 ovzaf	3693 owkof	3743 owxer	3793 oxmaf	3843 oyerp	3893 oytux	3943 ozhaj	3993 ozval
3544 ovbus	3594 ovmab	3644 ovzed	3694 owkux	3744 owyez	3794 oxmed	3844 oyesh	3894 oyual	3944 ozhir	3994 ozven
3545 ovbyt	3595 ovmev	3645 ovzib	3695 owlak	3745 owyok	3795 oxmoc	3845 oyeuc	3895 oyubu	3945 ozhox	3995 ozvit
3546 ovcam	3596 ovmim	3646 ovzly	3696 owlev	3746 owyuc	3796 oxmuz	3846 oyfol	3896 oyude	3946 ozhuk	3996 ozvos
3547 ovcel	3597 ovmon	3647 ovzoc	3697 owlor	3747 owzac	3797 oxnox	3847 oyfra	3897 oyufi	3947 ozigo	3997 ozway
3548 ovcow	3598 ovmuk	3648 ovzuz	3698 owlys	3748 owzep	3798 oxodu	3848 oyfuf	3898 oyuor	3948 oziph	3998 ozwex
3549 ovcud	3599 ovnan	3649 owahi	3699 owmaw	3749 owzim	3799 oxogo	3849 oygaz	3899 oyvis	3949 ozizu	3999 ozxin

ozzan ozzik ozzum paafo

SECTION B

4000 paahy	**4050** palik	**4100** peand	**4150** peora	**4200** pigob	**4250** piuko	**4300** pohaf	**4350** potal	**4400** pugec	**4450** puses
4001 paang	4051 palof	4101 peary	4151 peows	4201 piguf	4251 piule	4301 poher	4351 poteg	4401 pugif	4451 pusiz
4002 paape	4052 paluc	4102 pebat	4152 peozo	4202 pihil	4252 piuna	4302 pohis	4352 potyx	4402 pugol	4452 pusof
4003 paara	4053 pamar	4103 pebir	4153 pepip	4203 pihow	4253 pivas	4303 pohod	4353 poucu	4403 pugub	4453 pusuc
4004 pabaf	4054 pamid	4104 pebow	4154 pepla	4204 pijal	4254 pivej	4304 pohuz	4354 pouki	4404 puhav	4454 pusyr
4005 pabed	4055 pamog	4105 pebul	4155 perah	4205 pijen	4255 pivib	4305 poidi	4355 poung	4405 puhit	4455 putaw
4006 pabit	4056 pamux	4106 pecal	4156 perog	4206 pijos	4256 pivor	4306 poifu	4356 poupa	4406 puhoy	4456 putex
4007 paboc	4057 pamys	4107 pecef	4157 pesak	4207 pijut	4257 piwah	4307 poilk	4357 pouso	4407 puhul	4457 putuy
4008 pabry	4058 panas	4108 pecim	4158 peshu	4208 pikam	4258 piwey	4308 poisa	4358 povad	4408 puibe	4458 puvar
4009 pabuz	4059 panet	4109 pecot	4159 pesme	4209 pikel	4259 piwho	4309 poite	4359 povec	4409 puica	4459 puvez
4010 pacag	**4060** panor	**4110** pedad	**4160** petek	**4210** pikif	**4260** pizab	**4310** poivy	**4360** povir	**4410** puifo	**4460** puvin
4011 paceb	4061 panug	4111 pedec	4161 petho	4211 pikro	4261 pizef	4311 pojan	4361 povol	4411 puigu	4461 puvog
4012 pacil	4062 panyx	4112 pedol	4162 petuz	4212 pikud	4262 pizit	4312 pojeb	4362 powac	4412 puimp	4462 puwat
4013 pacox	4063 paoga	4113 pedub	4163 pevur	4213 pilan	4263 poaby	4313 pojik	4363 powed	4413 pujaf	4463 puwid
4014 pacun	4064 paolu	4114 pedym	4164 pewam	4214 pilet	4264 poadz	4314 pokap	4364 poyls	4414 pujed	4464 puyah
4015 pacyd	4065 paopi	4115 peeda	4165 pewer	4215 pimap	4265 poaho	4315 pokes	4365 poyoc	4415 pujoc	4465 puyen
4016 paday	4066 paovo	4116 peemb	4166 peyap	4216 pimes	4266 poalc	4316 pokim	4366 pozat	4416 pukag	4466 puyow
4017 padef	4067 paoxe	4117 peeni	4167 peyel	4217 pimof	4267 poari	4317 pokuc	4367 pozom	4417 pukeb	4467 puyut
4018 padoz	4068 papav	4118 peerz	4168 pezas	4218 pimuc	4268 poasm	4318 polez	4368 puaba	4418 pukil	4468 puzet
4019 paduv	4069 papex	4119 peetu	4169 pezor	4219 pimyt	4269 poaxu	4319 polid	4369 puach	4419 pukun	4469 puzop
4020 paect	**4070** papli	**4120** peexo	**4170** pezuc	**4220** pinac	**4270** pobag	**4320** polog	**4370** puaju	**4420** pukyd	**4470** puzur
4021 paegu	4071 papop	4121 peeye	4171 piagy	4221 pineg	4271 pobef	4321 polux	4371 pualy	4421 pulay	4471 pyacy
4022 paeko	4072 papur	4122 pefed	4172 piala	4222 piniv	4272 pobib	4322 polym	4372 puant	4422 pulef	4472 pyadu
4023 paems	4073 papyk	4123 pefis	4173 piawl	4223 pinom	4273 pobox	4323 pomas	4373 puase	4423 pulig	4473 pyafa
4024 paepy	4074 paraz	4124 pegaw	4174 pibad	4224 pinys	4274 pobun	4324 pomek	4374 puazo	4424 puloz	4474 pyale
4025 paexa	4075 parep	4125 pegig	4175 pibec	4225 pioce	4275 pobyl	4325 pomit	4375 pubax	4425 puluv	4475 pyani
4026 paezi	4076 parif	4126 pegro	4176 pibik	4226 piodo	4276 pocay	4326 pomor	4376 publo	4426 pumak	4476 pyaph
4027 pafah	4077 parok	4127 peguy	4177 pibol	4227 pioli	4277 pocet	4327 pomug	4377 pubup	4427 pumem	4477 pyarx
4028 pafem	4078 parud	4128 pehan	4178 pibub	4228 pioms	4278 pocoz	4328 ponav	4378 pubyn	4428 pumip	4478 pyato
4029 pafyl	4079 paryc	4129 pehoc	4179 picaf	4229 pionz	4279 podak	4329 poneh	4379 pucaz	4429 pumob	4479 pybak
4030 pagal	**4080** pasev	**4130** pehyl	**4180** piced	**4230** pioth	**4280** podem	**4330** ponop	**4380** pucep	**4430** pumuf	**4480** pybem
4031 pagen	4081 pason	4131 peims	4181 pichi	4231 piowa	4281 podig	4331 ponur	4381 pucro	4431 pumyg	4481 pybid
4032 pagiv	4082 pasym	4132 peirp	4182 picuz	4232 pioxu	4282 podob	4332 poobe	4382 pucus	4432 punek	4482 pybob
4033 pagut	4083 patec	4133 peitz	4183 picyk	4233 pipic	4283 poduf	4333 poock	4383 pudab	4433 punix	4483 pybuf
4034 paheh	4084 patub	4134 pejav	4184 pidag	4234 pipov	4284 podyn	4334 poomu	4384 pudev	4434 punyo	4484 pycaw
4035 pahin	4085 patyp	4135 pejod	4185 pideb	4235 pirav	4285 poebo	4335 pooni	4385 pudic	4435 puobs	4485 pycir
4036 pahuy	4086 paubi	4136 pejum	4186 pidid	4236 pirex	4286 poede	4336 poopy	4386 pudon	4436 puoda	4486 pycoy
4037 paidu	4087 pauoy	4137 pekay	4187 pidox	4237 pirip	4287 poefy	4337 poota	4387 pudry	4437 puofi	4487 pycut
4038 paily	4088 pauge	4138 pekex	4188 pidun	4238 pirot	4288 poeku	4338 popif	4388 puduk	4438 puojo	4488 pydac
4039 paino	4089 paups	4139 peklo	4189 pieky	4239 pirul	4289 poela	4339 porax	4389 puebu	4439 puoku	4489 pydeg
4040 paiva	**4090** pauxu	**4140** pekut	**4190** pield	**4240** pirym	**4290** poern	**4340** porhe	**4390** puele	**4440** puowl	**4490** pydin
4041 paize	4091 pauza	4141 pelab	4191 piepi	4241 pisez	4291 pofaw	4341 porov	4391 puens	4441 pupam	4491 pydul
4042 pajig	4092 pavac	4142 peley	4192 pieru	4242 pisin	4292 pofen	4342 porup	4392 pueph	4442 pupel	4492 pydyp
4043 pajus	4093 paveg	4143 pelil	4193 pifat	4243 pisog	4293 pofic	4343 poryd	4393 puerd	4443 pupiv	4493 pyefi
4044 pakan	4094 pavom	4144 pelop	4194 pifeh	4244 pisux	4294 pofoy	4344 posaz	4394 puewi	4444 pupot	4494 pyega
4045 pakey	4095 pawak	4145 peluk	4195 pifim	4245 pitaz	4295 pofut	4345 posep	4395 pufac	4445 pupud	4495 pyeke
4046 pakip	4096 pawic	4146 pemew	4196 pifoz	4246 pitix	4296 pogam	4346 posmi	4396 pufeg	4446 puran	4496 pyelm
4047 pakod	4097 pawol	4147 pemoy	4197 pigak	4247 pityr	4297 pogel	4347 posow	4397 pufib	4447 purew	4497 pyepu
4048 pakum	4098 pazax	4148 penin	4198 pigem	4248 piudi	4298 pogip	4348 posty	4398 pufom	4448 purod	4498 pyery
4049 palap	4099 pazer	4149 penyt	4199 pigly	4249 piugu	4299 pogot	4349 posus	4399 pufru	4449 pusap	4499 pyeth

4500 pyfam	**4550** pypok	**4600** raere	**4650** rapeb	**4700** regit	**4750** revaz	**4800** rigif	**4850** riutu	**4900** roidy	**4950** rousa
4501 pyfel	4551 pypus	4601 raeso	4651 rapow	4701 regor	4751 revey	4801 rigub	4851 rivar	4901 roips	4951 roveh
4502 pyfig	4552 pyrab	4602 raevu	4652 rapun	4702 regug	4752 revim	4802 rihav	4852 rivez	4902 roire	4952 rovip
4503 pyfot	4553 pyrev	4603 rafav	4653 raraw	4703 rehec	4753 revok	4803 rihoy	4853 rivin	4903 roita	4953 rovow
4504 pyfud	4554 pyriz	4604 rafey	4654 raref	4704 reids	4754 revuf	4804 rihul	4854 rivog	4904 roker	4954 rovus
4505 pygan	4555 pyron	4605 rafiz	4655 rasah	4705 reiga	4755 rewag	4805 rijaf	4855 riwat	4905 rokig	4955 rovyt
4506 pygew	4556 pyruk	4606 rafop	4656 rasem	4706 reinu	4756 reweb	4806 rijed	4856 riwid	4906 rokup	4956 rowav
4507 pygic	4557 pysad	4607 rafur	4657 rasip	4707 reipy	4757 rewow	4807 rijoc	4857 rizet	4907 rokyf	4957 rowec
4508 pygod	4558 pysec	4608 ragax	4658 rasob	4708 reixo	4758 rewry	4808 rikag	4858 rizop	4908 rolem	4958 rowon
4509 pygum	4559 pysol	4609 ragib	4659 ratep	4709 rejaw	4759 rexal	4809 rikeb	4859 rizur	4909 rolin	4959 roxol
4510 pyhay	**4560** pyspy	**4610** ragov	**4660** rativ	**4710** rekef	**4760** reyaf	**4810** rikil	**4860** roafy	**4910** rolob	**4960** royba
4511 pyhet	4561 pysti	4611 ragup	4661 ratul	4711 rekoz	4761 reyed	4811 rikun	4861 roals	4911 roluf	4961 rozle
4512 pyhor	4562 pysub	4612 rahaj	4662 raudy	4712 rekru	4762 reyop	4812 rikyd	4862 roamu	4912 rolyx	4962 rozot
4513 pyhun	4563 pyted	4613 rahir	4663 raval	4713 relah	4763 rezad	4813 rilay	4863 roank	4913 romos	4963 ruaci
4514 pyhym	4564 pytip	4614 rahox	4664 ravit	4714 relen	4764 rezig	4814 rilef	4864 roapi	4914 romyr	4964 ruafu
4515 pyici	4565 pytox	4615 rahuk	4665 ravos	4715 relos	4765 rezol	4815 rilig	4865 roart	4915 ronam	4965 ruaga
4516 pyiku	4566 pytuc	4616 rahys	4666 raway	4716 reluw	4766 rezub	4816 riloz	4866 roavo	4916 ronel	4966 rualk
4517 pyilo	4567 pytyt	4617 raics	4667 rawex	4717 remam	4767 riaba	4817 riluv	4867 roawa	4917 ronit	4967 ruash
4518 pyima	4568 pyubo	4618 raigo	4668 raxin	4718 remel	4768 riach	4818 rimak	4868 roaxe	4918 ronud	4968 rubam
4519 pyise	4569 pyuka	4619 raika	4669 razan	4719 remot	4769 riaju	4819 rimem	4869 robaj	4919 ronyc	4969 rubiv
4520 pykar	**4570** pyulu	**4620** raiph	**4670** razik	**4720** remud	**4770** rialy	**4820** rimip	**4870** robil	**4920** roocu	**4970** ruboy
4521 pykib	4571 pyune	4621 raity	4671 razum	4721 renar	4771 riant	4821 rimuf	4871 roboz	4921 rooko	4971 rubud
4522 pykog	4572 pyvat	4622 raizu	4672 reacu	4722 renez	4772 riase	4822 rimyg	4872 robys	4922 rooma	4972 rucac
4523 pylas	4573 pyvix	4623 rajat	4673 reaxy	4723 renip	4773 riazo	4823 rinal	4873 rocan	4923 roope	4973 ruceg
4524 pylek	4574 pyvoc	4624 rajew	4674 reazi	4724 renog	4774 riber	4824 rinek	4874 rocey	4924 roosy	4974 ruclu
4525 pylim	4575 pywal	4625 rakab	4675 rebab	4725 renux	4775 riblo	4825 rinix	4875 rocod	4925 rooti	4975 rucom
4526 pylow	4576 pywiv	4626 rakev	4676 rebev	4726 renyk	4776 ribyn	4826 rinyo	4876 rocum	4926 ropas	4976 rudan
4527 pylug	4577 raahs	4627 rakis	4677 rebon	4727 reoco	4777 ricaz	4827 riobs	4877 rodaw	4927 ropek	4977 rudey
4528 pylyb	4578 raaki	4628 raklu	4678 rebuk	4728 reohm	4778 ricep	4828 rioda	4878 rodex	4928 ropix	4978 rudif
4529 pymaj	4579 raalb	4629 rakon	4679 reciz	4729 reoki	4779 ricro	4829 riofi	4879 rodis	4929 ropor	4979 rudod
4530 pymeh	**4580** raamt	**4630** rakyr	**4680** recob	**4730** reolt	**4780** ricus	**4830** riojo	**4880** rodly	**4930** ropug	**4980** rudut
4531 pymik	4581 raawe	4631 ralac	4681 recra	4731 reowe	4781 ridev	4831 rioku	4881 roejo	4931 rorap	4981 ruecu
4532 pymop	4582 raaya	4632 raleg	4682 recyc	4732 repan	4782 ridic	4832 riowl	4882 roelf	4932 rores	4982 rueja
4533 pymur	4583 rabap	4633 ralif	4683 redap	4733 repet	4783 ridon	4833 ripel	4883 roene	4933 roriv	4983 rueli
4534 pynax	4584 rabes	4634 ralom	4684 redes	4734 rephy	4784 ridry	4834 ripiv	4884 roeva	4934 roruc	4984 ruerx
4535 pyner	4585 rabig	4635 ralyp	4685 redof	4735 repib	4785 riduk	4835 ripud	4885 roexu	4935 roryb	4985 rueve
4536 pynif	4586 rabof	4636 ramec	4686 reduc	4736 repum	4786 riebu	4836 rirew	4886 rofar	4936 rosac	4986 rufap
4537 pynov	4587 rabuc	4637 ramix	4687 reeha	4737 rerac	4787 riefa	4837 ririk	4887 rofez	4937 roseg	4987 rufes
4538 pynup	4588 rabyb	4638 ramol	4688 reeny	4738 rereg	4788 riele	4838 riryl	4888 rofik	4938 rosom	4988 rufit
4539 pynyg	4589 racar	4639 ranaf	4689 reepo	4739 rerom	4789 riens	4839 risiz	4889 rofog	4939 rosuk	4989 rufro
4540 pyoch	**4590** rachy	**4640** raned	**4690** reets	**4740** rerus	**4790** rieph	**4840** risof	**4890** rofux	**4940** rosyn	**4990** ruful
4541 pyode	4591 racic	4641 ranoc	4691 reexi	4741 resat	4791 rierd	4841 risyr	4891 rogaf	4941 rotak	4991 rugar
4542 pyoja	4592 racog	4642 ranuz	4692 reeze	4742 resoy	4792 rieto	4842 ritaw	4892 roged	4942 rotet	4992 rugez
4543 pyomi	4593 racux	4643 raoca	4693 refay	4743 resul	4793 riewi	4843 ritex	4893 rogno	4943 rotij	4993 rugog
4544 pyops	4594 radek	4644 raodi	4694 refex	4744 reter	4794 rifac	4844 rithi	4894 rogry	4944 rotoc	4994 ruhat
4545 pyorf	4595 radug	4645 raoho	4695 refil	4745 retid	4795 rifeg	4845 ritor	4895 roguz	4945 rotun	4995 ruhib
4546 pyotu	4596 radyc	4646 raonu	4696 refoh	4746 retys	4796 rifib	4846 rituy	4896 rohax	4946 rotyd	4996 ruhov
4547 pypaz	4597 raelz	4647 raost	4697 refuz	4747 reufu	4797 rifom	4847 riuga	4897 rohic	4947 rounu	4997 ruhuf
4548 pypep	4598 raemy	4648 raoxy	4698 regas	4748 reujo	4798 rifru	4848 riumi	4898 rohub	4948 roupo	4998 ruhyg
4549 pypit	4599 raept	4649 rapag	4699 regek	4749 reulk	4799 rigec	4849 riurn	4899 roick	4949 rourb	4999 ruiba

ruife ruimi ruiru ruiso

SECTION B

5000 rujas	**5050** ruwis	**5100** rylil	**5150** sadro	**5200** sanep	**5250** searx	**5300** selug	**5350** siamt	**5400** sioca	**5450** soerg
5001 rujet	5051 ruyak	5101 rylop	5151 sadud	5201 sanim	5251 sebak	5301 selyb	5351 siawe	5401 siodi	5451 soese
5002 rujor	5052 ruzal	5102 ryluk	5152 sadyt	5202 sanok	5252 sebid	5302 semaj	5352 siaya	5402 sioho	5452 sogav
5003 rukav	5053 ruzen	5103 rymew	5153 saebe	5203 saofo	5253 sebob	5303 semik	5353 sibap	5403 sionu	5453 sogim
5004 rukew	5054 ruzip	5104 rymoy	5154 saecs	5204 saoke	5254 sebuf	5304 semop	5354 sibes	5404 siost	5454 sogop
5005 rukiz	5055 ruzos	5105 rynin	5155 saeju	5205 saold	5255 sebyx	5305 senax	5355 sibig	5405 sioxy	5455 sohab
5006 rukry	5056 ruzug	5106 rynos	5156 saend	5206 saont	5256 secaw	5306 sener	5356 sibof	5406 sipag	5456 sohek
5007 rulax	5057 ryait	5107 rynyt	5157 saerf	5207 saopu	5257 secir	5307 senif	5357 sibuc	5407 sipeb	5457 sohig
5008 rulot	5058 ryand	5108 ryoaf	5158 saewo	5208 saord	5258 secoy	5308 senov	5358 sibyb	5408 sipow	5458 sohon
5009 rulup	5059 ryary	5109 ryoen	5159 safan	5209 saovi	5259 secut	5309 senup	5359 sicar	5409 sipun	5459 sohup
5010 rumaz	**5060** rybat	**5110** ryoid	**5160** safew	**5210** saoxa	**5260** sedac	**5310** senyg	**5360** sichy	**5410** siraw	**5460** soimo
5011 rumep	5061 rybeb	5111 ryora	5161 safod	5211 sapac	5261 sedeg	5311 seoch	5361 sicic	5411 siref	5461 soiny
5012 rumin	5062 rybir	5112 ryoux	5162 safri	5212 sapeg	5262 sedin	5312 seode	5362 sicog	5412 siroz	5462 soipu
5013 rumok	5063 rybow	5113 ryows	5163 safum	5213 saplu	5263 sedul	5313 seoja	5363 sicux	5413 sisah	5463 soiwa
5014 runab	5064 rybul	5114 ryozo	5164 safys	5214 sapom	5264 sedyp	5314 seomi	5364 sidas	5414 sisem	5464 sojah
5015 runev	5065 rycal	5115 rypip	5165 sagap	5215 sapyx	5265 seefi	5315 seops	5365 sidor	5415 sisip	5465 sojer
5016 runig	5066 rycef	5116 rypla	5166 sages	5216 sarad	5266 seega	5316 seorf	5366 sidyc	5416 sisob	5466 sojis
5017 runon	5067 rycim	5117 rypof	5167 sagiz	5217 sarec	5267 seeke	5317 seotu	5367 sielz	5417 sisuf	5467 sokaz
5018 runuk	5068 rycot	5118 rypre	5168 sagof	5218 sarol	5268 seelm	5318 sepaz	5368 siemy	5418 sitam	5468 sokep
5019 runyp	5069 rydad	5119 ryrah	5169 saguc	5219 sarub	5269 seepu	5319 sepep	5369 siept	5419 sitep	5469 sokok
5020 ruody	**5070** rydec	**5120** ryrog	**5170** sagym	**5220** sasaf	**5270** seery	**5320** sepit	**5370** sievu	**5420** sitiv	**5470** sokus
5021 ruola	5071 rydol	5121 rysak	5171 sahak	5221 sased	5271 seeth	5321 sepok	5371 sifav	5421 situl	5471 sokym
5022 ruomo	5072 rydub	5122 ryshu	5172 sahey	5222 sashy	5272 sefam	5322 sepus	5372 sifey	5422 siudy	5472 solal
5023 ruong	5073 rydym	5123 rysme	5173 sahip	5223 sasig	5273 sefel	5323 serab	5373 sifiz	5423 sival	5473 solib
5024 ruose	5074 ryeda	5124 rytag	5174 sahoz	5224 sasoc	5274 sefig	5324 serev	5374 sifop	5424 siven	5474 soloh
5025 rupad	5075 ryemb	5125 rytek	5175 sahux	5225 sasuz	5275 sefot	5325 seriz	5375 sifur	5425 sivit	5475 solyr
5026 rupec	5076 ryeni	5126 rytho	5176 saich	5226 satav	5276 sefud	5326 seruk	5376 sigib	5426 sivos	5476 someg
5027 rupij	5077 ryeps	5127 rytuz	5177 saido	5227 satem	5277 segan	5327 sesad	5377 sigov	5427 siway	5477 somif
5028 rupol	5078 ryerz	5128 ryupu	5178 saine	5228 satuf	5278 segew	5328 sesec	5378 sihaj	5428 siwex	5478 somom
5029 rupub	5079 ryetu	5129 ryusi	5179 saiza	5229 saubs	5279 segic	5329 sesol	5379 sihir	5429 sixin	5479 sonol
5030 rupyr	**5080** ryexo	**5130** ryvur	**5180** sajez	**5230** sauck	**5280** segod	**5330** sespy	**5380** sihox	**5430** sizan	**5480** soobo
5031 ruraf	5081 ryfed	5131 rywam	5181 sajog	5231 sauly	5281 segum	5331 sesub	5381 sihuk	5431 sizum	5481 soofy
5032 rured	5082 ryfis	5132 rywer	5182 sakas	5232 sauno	5282 sehay	5332 seted	5382 sihys	5432 soagi	5482 soome
5033 ruroc	5083 rygaw	5133 rywiz	5183 sakek	5233 sauzi	5283 sehet	5333 setip	5383 sijat	5433 soamy	5483 soonc
5034 ruruz	5084 ryget	5134 ryxem	5184 sakid	5234 savag	5284 sehor	5334 setox	5384 sijew	5434 soarn	5484 soopa
5035 rusag	5085 rygig	5135 saaco	5185 sakor	5235 savil	5285 sehym	5335 setuc	5385 sikev	5435 soawo	5485 soori
5036 ruseb	5086 rygro	5136 saada	5186 sakug	5236 savox	5286 seici	5336 setyt	5386 siklu	5436 sobiz	5486 soots
5037 rusik	5087 ryguy	5137 saame	5187 sakyp	5237 savun	5287 seigy	5337 seubo	5387 sikon	5437 sobod	5487 sopaf
5038 rusox	5088 ryhan	5138 saans	5188 salaw	5238 sawir	5288 seiku	5338 seuka	5388 sikyr	5438 sobum	5488 sopil
5039 rusun	5089 ryhoc	5139 saarg	5189 saleh	5239 sawob	5289 seilo	5339 seulu	5389 silac	5439 sobyc	5489 sopoc
5040 rusyd	**5090** ryiha	**5140** saavu	**5190** salit	**5240** sayet	**5290** sejap	**5340** seune	**5390** sileg	**5440** soces	**5490** sopuz
5041 rutah	5091 ryims	5141 saben	5191 salur	5241 sayup	5291 sejes	5341 sevat	5391 silif	5441 sociv	5491 soreb
5042 rutob	5092 ryion	5142 sabix	5192 salyd	5242 sazic	5292 sejil	5342 sevix	5392 silom	5442 socof	5492 sorox
5043 rutyn	5093 ryirp	5143 sabos	5193 samax	5243 sazuk	5293 sejof	5343 sevoc	5393 silyp	5443 socuc	5493 soryf
5044 ruvay	5094 ryitz	5144 sacab	5194 samer	5244 seacy	5294 sejux	5344 sewal	5394 simad	5444 socyx	5494 sosys
5045 ruvef	5095 rykay	5145 sacli	5195 samib	5245 seadu	5295 sekar	5345 sewiv	5395 simec	5445 sodez	5495 sotha
5046 ruvid	5096 rykex	5146 sacon	5196 samov	5246 seafa	5296 sekib	5346 seyom	5396 sinaf	5446 sodux	5496 souba
5047 ruvoz	5097 ryklo	5147 sadam	5197 samut	5247 seale	5297 sekog	5347 siahs	5397 sined	5447 soedy	5497 soufe
5048 ruwaw	5098 rykut	5148 sadel	5198 samyn	5248 seani	5298 selek	5348 siaki	5398 sinoc	5448 soefo	5498 souku
5049 ruwem	5099 rylab	5149 sadiv	5199 sanaz	5249 seaph	5299 selow	5349 sialb	5399 sinuz	5449 soelv	5499 souni

SECTION B

5500 souzo	**5550** suinz	**5600** suwor	**5650** syitu	**5700** syvif	**5750** tajef	**5800** tawyn	**5850** teino	**5900** teulo	**5950** tihuz
5501 sovak	5551 suism	5601 suxel	5651 syive	5701 syvoy	5751 tajid	5801 tayca	5851 teiva	5901 teups	5951 tijan
5502 sovem	5552 sujeh	5602 suyar	5652 sykak	5702 sywax	5752 tajot	5802 tayem	5852 teize	5902 teuza	5952 tijeb
5503 sovic	5553 sujim	5603 suyex	5653 sykeh	5703 sywig	5753 tajup	5803 tayux	5853 tejel	5903 tevac	5953 tikap
5504 sovul	5554 sujoy	5604 suyog	5654 sykob	5704 sywom	5754 takak	5804 tazaf	5854 tejig	5904 teveg	5954 tikes
5505 soweh	5555 sujub	5605 suzek	5655 sykuf	5705 taade	5755 takeh	5805 tazed	5855 tekan	5905 tevom	5955 tikim
5506 soyle	5556 sukaf	5606 suzir	5656 sykys	5706 taaku	5756 takob	5806 tazib	5856 tekey	5906 tewak	5956 tikuc
5507 soyub	5557 suked	5607 syade	5657 sylov	5707 taamo	5757 takuf	5807 tazly	5857 tekip	5907 tewic	5957 tilar
5508 sozam	5558 sukli	5608 syaku	5658 sylul	5708 taany	5758 takys	5808 tazoc	5858 tekod	5908 tewol	5958 tilez
5509 sozel	5559 sukoc	5609 syamo	5659 sylyg	5709 taapt	5759 talat	5809 tazuz	5859 tekum	5909 tezer	5959 tilid
5510 sozoy	**5560** sulag	**5610** syany	**5660** symab	**5710** taars	**5760** talov	**5810** teafo	**5860** telap	**5910** tiaby	**5960** tilog
5511 suabi	5561 suleb	5611 syapt	5661 symev	5711 taava	5761 talul	5811 teahy	5861 teles	5911 tiadz	5961 tilux
5512 suact	5562 suliz	5612 syars	5662 symim	5712 tabin	5762 talyg	5812 teang	5862 telik	5912 tiaho	5962 tilym
5513 suady	5563 sulox	5613 syava	5663 symon	5713 tabok	5763 tamev	5813 teape	5863 telof	5913 tialc	5963 timas
5514 suama	5564 sulun	5614 sybaz	5664 symuk	5714 tabre	5764 tamon	5814 teara	5864 teluc	5914 tiana	5964 timek
5515 suapo	5565 sulyt	5615 sybin	5665 synan	5715 tabyt	5765 tamuk	5815 tebaf	5865 temar	5915 tiari	5965 timit
5516 subav	5566 sumah	5616 sybok	5666 syney	5716 tacam	5766 tanan	5816 tebed	5866 temez	5916 tiasm	5966 timug
5517 subij	5567 sumef	5617 sybre	5667 synum	5717 tacel	5767 taney	5817 tebit	5867 temid	5917 tiaxu	5967 tinav
5518 suble	5568 sumic	5618 sybyt	5668 syogi	5718 tacis	5768 tanod	5818 teboc	5868 temog	5918 tibag	5968 tineh
5519 suboh	5569 sumoz	5619 sycam	5669 syomy	5719 tacow	5769 taogi	5819 tebry	5869 temux	5919 tibef	5969 tinop
5520 sucib	**5570** sunay	**5620** sycel	**5670** syono	**5720** tacud	**5770** taomy	**5820** tebuz	**5870** temys	**5920** tibib	**5970** tinur
5521 sucov	5571 sunem	5621 sycis	5671 syork	5721 tacyp	5771 taono	5821 tecag	5871 tenas	5921 tibox	5971 tiobe
5522 sucup	5572 sunid	5622 sycud	5672 syosa	5722 tadal	5772 taork	5822 teceb	5872 tenug	5922 tibun	5972 tiock
5523 sudaz	5573 sunuf	5623 sycyp	5673 syovu	5723 taden	5773 taosa	5823 tecil	5873 tenyx	5923 tibyl	5973 tiolo
5524 sudep	5574 suoak	5624 sydal	5674 sypap	5724 tadik	5774 taovu	5824 tecox	5874 teoga	5924 ticet	5974 tiomu
5525 sudix	5575 suofu	5625 syden	5675 sypes	5725 tados	5775 tapap	5825 tecun	5875 teolu	5925 tidak	5975 tioni
5526 sudok	5576 suoly	5626 sydik	5676 sypiz	5726 taduy	5776 tapro	5826 tecyd	5876 teopi	5926 tidem	5976 tiopy
5527 sudus	5577 suomp	5627 sydos	5677 sypro	5727 taecy	5777 tapuc	5827 teday	5877 teovo	5927 tidig	5977 tiota
5528 suedi	5578 suone	5628 syduy	5678 sypuc	5728 taeka	5778 tapyd	5828 tedef	5878 teoxe	5928 tidob	5978 tipev
5529 sueng	5579 suort	5629 syeka	5679 sypyd	5729 taelb	5779 taryx	5829 tedoz	5879 tepav	5929 tiduf	5979 tipif
5530 suerl	**5580** suosi	**5630** syelb	**5680** syrar	**5730** taenu	**5780** tasas	**5830** teduv	**5880** tepex	**5930** tidyn	**5980** tipon
5531 suesy	5581 suowo	5631 syenu	5681 syrit	5731 taesi	5781 tasek	5831 teect	5881 tepur	5931 tiebo	5981 tipuk
5532 suewa	5582 supal	5632 syesi	5682 syryx	5732 tafad	5782 tasor	5832 teegu	5882 tepyk	5932 tiefy	5982 tipyc
5533 suexe	5583 supen	5633 syfad	5683 sysas	5733 tafec	5783 taspi	5833 teeko	5883 teraz	5933 tieku	5983 tirov
5534 sufab	5584 supig	5634 syfec	5684 sysek	5734 taflo	5784 tasug	5834 teepy	5884 terep	5934 tiela	5984 tiryd
5535 sufev	5585 surez	5635 syflo	5685 sysor	5735 tafry	5785 tasyl	5835 teexa	5885 terif	5935 tiern	5985 tisaz
5536 sufip	5586 surhi	5636 syfry	5686 syspi	5736 tafub	5786 tatil	5836 teezi	5886 terok	5936 tieti	5986 tisep
5537 suflu	5587 surow	5637 syfub	5687 sysug	5737 tagag	5787 tatof	5837 tefah	5887 terud	5937 tifaw	5987 tismi
5538 sufon	5588 suryn	5638 sygag	5688 sysyl	5738 tagir	5788 tatye	5838 tefem	5888 teryc	5938 tifen	5988 tisow
5539 sugac	5589 susew	5639 sygir	5689 sytac	5739 tagle	5789 taugo	5839 tefyl	5889 tesab	5939 tific	5989 tisty
5540 sugeg	**5590** susil	**5640** sygle	**5690** sytil	**5740** tagun	**5790** tauli	**5840** tegal	**5890** tesev	**5940** tifoy	**5990** tital
5541 sugin	5591 sutas	5641 sygun	5691 sytof	5741 tahah	5791 taumu	5841 tegen	5891 tesis	5941 tifut	5991 titeg
5542 sugom	5592 sutey	5642 syhah	5692 sytye	5742 taheg	5792 taush	5842 tegiv	5892 tesup	5942 tigam	5992 titin
5543 sugyl	5593 sutik	5643 syheg	5693 syugo	5743 tahol	5793 tauta	5843 tegos	5893 tesym	5943 tigel	5993 titum
5544 suhad	5594 sutol	5644 syhol	5694 syuli	5744 taiby	5794 tavav	5844 tegut	5894 tetec	5944 tigip	5994 tityx
5545 suhif	5595 sutut	5645 syhur	5695 syumu	5745 taico	5795 tavet	5845 teheh	5895 tetir	5945 tigot	5995 tiucu
5546 suhot	5596 suvap	5646 syiby	5696 syush	5746 taitu	5796 tavif	5846 tehin	5896 tetub	5946 tigru	5996 tiuki
5547 suhug	5597 suves	5647 syico	5697 syuta	5747 taive	5797 tavoy	5847 tehuy	5897 tetyp	5947 tiher	5997 tiung
5548 suhyr	5598 suviv	5648 syifa	5698 syvav	5748 taizi	5798 tawig	5848 teidu	5898 teucy	5948 tihis	5998 tiupa
5549 suicy	5599 suwit	5649 syils	5699 syvet	5749 tajay	5799 tawom	5849 teily	5899 teuge	5949 tihod	5999 tiuso

tivad tiveo tivir tivol

[redacted]

SECTION B

6000 tiwac	**6050** toreg	**6100** tuhez	**6150** tutux	**6200** tygno	**6250** tytoc	**6300** ubhif	**6350** ubsod	**6400** ucfeh	**6450** ucper
6001 tiwed	6051 torom	6101 tuhid	6151 tuvis	6201 tygry	6251 tytun	6301 ubhot	6351 ubsum	6401 ucfim	6451 ucpic
6002 tizat	6052 tosul	6102 tuhob	6152 tuvon	6202 tyguz	6252 tyugi	6302 ubhug	6352 ubtas	6402 ucgak	6452 ucpov
6003 toacu	6053 totys	6103 tuimy	6153 tuwaf	6203 tyhax	6253 tyunu	6303 ubhyr	6353 ubtey	6403 ucgem	6453 ucpup
6004 toams	6054 touby	6104 tuini	6154 tuwel	6204 tyhic	6254 tyupo	6304 ubicy	6354 ubtol	6404 ucgly	6454 ucrav
6005 toaro	6055 toufu	6105 tuisk	6155 tuwip	6205 tyhub	6255 tyurb	6305 ubigi	6355 ubtut	6405 ucgob	6455 ucrex
6006 toata	6056 toujo	6106 tuizo	6156 tuwog	6206 tyick	6256 tyusa	6306 ubinz	6356 ubufa	6406 ucguf	6456 ucrip
6007 toave	6057 toulk	6107 tujad	6157 tuwye	6207 tyidy	6257 tyute	6307 ubiro	6357 ubunc	6407 uchow	6457 ucrot
6008 toaxy	6058 tovaz	6108 tujek	6158 tuxar	6208 tyifi	6258 tyveh	6308 ubism	6358 uburg	6408 ucibu	6458 ucrym
6009 toazi	6059 tovey	6109 tujow	6159 tuxen	6209 tyips	6259 tyvip	6309 ubjeh	6359 ubust	6409 ucify	6459 ucsar
6010 tobab	**6060** tovok	**6110** tujur	**6160** tuxim	**6210** tyire	**6260** tyvow	**6310** ubjim	**6360** ubuxo	**6410** ucige	**6460** ucsez
6011 tobev	6061 tovuf	6111 tukaw	6161 tuyef	6211 tyita	6261 tyvus	6311 ubjoy	6361 ubuze	6411 ucind	6461 ucsin
6012 tobuk	6062 toxal	6112 tukem	6162 tuygo	6212 tyker	6262 tywav	6312 ubjub	6362 ubvap	6412 ucirk	6462 ucsog
6013 tociz	6063 toxur	6113 tukot	6163 tuyne	6213 tykig	6263 tywec	6313 ubkaf	6363 ubves	6413 ucisl	6463 ucsux
6014 tocra	6064 toyaf	6114 tukuk	6164 tuyum	6214 tykup	6264 tywon	6314 ubked	6364 ubviv	6414 ucivo	6464 uctaz
6015 tocyc	6065 toyop	6115 tulav	6165 tuzap	6215 tylem	6265 tyxol	6315 ubkli	6365 ubwit	6415 ucixa	6465 uctix
6016 todof	6066 tozad	6116 tulep	6166 tuzif	6216 tylin	6266 ubact	6316 ubkoc	6366 ubwor	6416 ucjal	6466 uctus
6017 toduc	6067 tozig	6117 tulir	6167 tyafy	6217 tylob	6267 ubady	6317 ublag	6367 ubxel	6417 ucjen	6467 uctyr
6018 toeha	6068 tozol	6118 tulok	6168 tyals	6218 tyluf	6268 ubage	6318 ubleb	6368 ubyar	6418 ucjos	6468 ucudi
6019 toeny	6069 tozub	6119 tulyc	6169 tyamu	6219 tylyx	6269 ubama	6319 ubliz	6369 ubyex	6419 ucjut	6469 ucugu
6020 toepo	**6070** tuabs	**6120** tumat	**6170** tyank	**6220** tymal	**6270** ubapo	**6320** ublox	**6370** ubyog	**6420** uckam	**6470** ucuko
6021 toexi	6071 tuajo	6121 tumey	6171 tyapi	6221 tymen	6271 ubaux	6321 ublun	6371 ubzek	6421 uckel	6471 ucule
6022 tofoh	6072 tuake	6122 tumod	6172 tyart	6222 tymos	6272 ubbav	6322 ublyt	6372 ubzir	6422 uckif	6472 ucuna
6023 tofuz	6073 tualu	6123 tumup	6173 tyavo	6223 tymyr	6273 ubbij	6323 ubmah	6373 ucagy	6423 uckro	6473 ucvas
6024 togug	6074 tuarc	6124 tunak	6174 tyawa	6224 tynam	6274 ubble	6324 ubmef	6374 ucala	6424 uckud	6474 ucvej
6025 tohec	6075 tuavi	6125 tunec	6175 tyaxe	6225 tynel	6275 ubboh	6325 ubmic	6375 ucano	6425 uclan	6475 ucvib
6026 tohiv	6076 tubex	6126 tuniz	6176 tybaj	6226 tynit	6276 ubbur	6326 ubmoz	6376 ucare	6426 uclet	6476 ucvor
6027 toids	6077 tucah	6127 tunof	6177 tybil	6227 tynud	6277 ubcax	6327 ubnay	6377 ucasi	6427 uclis	6477 ucwah
6028 toiga	6078 tucig	6128 tunut	6178 tyboz	6228 tynyc	6278 ubcer	6328 ubnem	6378 ucatu	6428 uclod	6478 ucwey
6029 toime	6079 tucul	6129 tuoby	6179 tybys	6229 tyocu	6279 ubcib	6329 ubnob	6379 ucawl	6429 uclum	6479 ucwho
6030 toinu	**6080** tucym	**6130** tuocs	**6180** tycan	**6230** tyoko	**6280** ubcov	**6330** ubnuf	**6380** ucbad	**6430** ucmap	**6480** ucyon
6031 toiri	6081 tudet	6131 tuogn	6181 tycey	6231 tyoma	6281 ubcup	6331 uboak	6381 ucbec	6431 ucmes	6481 ucyra
6032 toixo	6082 tudom	6132 tuoka	6182 tycod	6232 tyope	6282 ubdaz	6332 ubofu	6382 ucbik	6432 ucmiz	6482 ucyug
6033 tokef	6083 tudyl	6133 tuole	6183 tycum	6233 tyoti	6283 ubdep	6333 uboly	6383 ucbol	6433 ucmuc	6483 uczab
6034 tokoz	6084 tuebi	6134 tuoro	6184 tydaw	6234 typas	6284 ubdix	6334 ubomp	6384 ucbub	6434 ucmyt	6484 udaco
6035 tokru	6085 tueft	6135 tuotz	6185 tydex	6235 typek	6285 ubdok	6335 ubone	6385 uccaf	6435 ucnac	6485 udada
6036 tolah	6086 tuelk	6136 tuoxi	6186 tydis	6236 typix	6286 ubedi	6336 ubort	6386 ucced	6436 ucneg	6486 udaik
6037 tolen	6087 tueme	6137 tupaj	6187 tydur	6237 typor	6287 ubemo	6337 ubosi	6387 uccoc	6437 ucniv	6487 udame
6038 tomel	6088 tuena	6138 tupho	6188 tyejo	6238 typug	6288 ubeng	6338 ubowo	6388 uccuz	6438 ucnom	6488 udans
6039 tomot	6089 tuerp	6139 tupuy	6189 tyelf	6239 tyrap	6289 uberl	6339 ubpal	6389 uccyk	6439 ucnys	6489 udarg
6040 tonar	**6090** tuesh	**6140** turay	**6190** tyene	**6240** tyriv	**6290** ubesy	**6340** ubpen	**6390** ucdag	**6440** ucoce	**6490** udavu
6041 tonez	6091 tufol	6141 turic	6191 tyeva	6241 tyruc	6291 ubewa	6341 ubpig	6391 ucdeb	6441 ucodo	6491 udbal
6042 tonyk	6092 tufra	6142 turoh	6192 tyexu	6242 tysac	6292 ubexe	6342 ubpos	6392 ucdid	6442 ucoli	6492 udben
6043 tooco	6093 tufuf	6143 tusko	6193 tyfar	6243 tyseg	6293 ubfab	6343 ubrat	6393 ucdox	6443 ucoms	6493 udbix
6044 tooki	6094 tugaz	6144 tusla	6194 tyfez	6244 tysir	6294 ubfev	6344 ubrez	6394 ucdun	6444 uconz	6494 udbos
6045 toolt	6095 tugli	6145 tustu	6195 tyfik	6245 tysom	6295 ubfip	6345 ubrhi	6395 uceky	6445 ucory	6495 udcab
6046 toons	6096 tugoc	6146 tuswe	6196 tyfog	6246 tysyn	6296 ubflu	6346 ubrow	6396 uceld	6446 ucoth	6496 udcli
6047 tophy	6097 tugre	6147 tusyf	6197 tyfux	6247 tytak	6297 ubfon	6347 ubryn	6397 ucepi	6447 ucowa	6497 udcon
6048 topib	6098 tugud	6148 tutan	6198 tygaf	6248 tytet	6298 ubgyl	6348 ubsew	6398 ucete	6448 ucoxu	6498 udcre
6049 torac	6099 tuhac	6149 tutop	6199 tyged	6249 tytij	6299 ubhad	6349 ubsil	6399 ucfat	6449 ucpax	6499 uddam

6500 uddel	**6550** udmyn	**6600** ufack	**6650** ufjah	**6700** ufufe	**6750** uggos	**6800** ugson	**6850** uhimy	**6900** uhwye	**6950** ujlax
6501 uddro	6551 udnaz	6601 ufaft	6651 ufjer	6701 ufuku	6751 uggut	6801 ugsup	6851 uhini	6901 uhxar	6951 ujler
6502 uddud	6552 udnep	6602 ufagi	6652 ufjis	6702 ufuni	6752 ugheh	6802 ugsym	6852 uhisk	6902 uhxim	6952 ujlup
6503 uddyt	6553 udnim	6603 ufahu	6653 ufkaz	6703 ufvak	6753 ughin	6803 ugtad	6853 uhizo	6903 uhzus	6953 ujmep
6504 udebe	6554 udnok	6604 ufaja	6654 ufkep	6704 ufvem	6754 ughuy	6804 ugtec	6854 uhjad	6904 ujaci	6954 ujmin
6505 udecs	6555 udnus	6605 ufamy	6655 ufkin	6705 ufvic	6755 ugily	6805 ugtir	6855 uhjek	6905 ujado	6955 ujmok
6506 udeju	6556 udofo	6606 ufarn	6656 ufkok	6706 ufvul	6756 ugino	6806 ugtub	6856 uhjow	6906 ujafu	6956 ujmus
6507 udend	6557 udoke	6607 ufasp	6657 ufkus	6707 ufweh	6757 ugiva	6807 ugtyp	6857 uhjur	6907 ujaga	6957 ujnig
6508 uderf	6558 udold	6608 ufawo	6658 ufkym	6708 ufyle	6758 ugize	6808 ugubi	6858 uhkaw	6908 ujalk	6958 ujnyp
6509 udeta	6559 udont	6609 ufban	6659 uflal	6709 ufyot	6759 ugjel	6809 ugucy	6859 uhkem	6909 ujash	6959 ujody
6510 udewo	**6560** udopu	**6610** ufbet	**6660** uflex	**6710** ufyub	**6760** ugjig	**6810** uguge	**6860** uhkot	**6910** ujbam	**6960** ujola
6511 udfan	6561 udorb	6611 ufbiz	6661 uflib	6711 ufzam	6761 ugjus	6811 ugulo	6861 uhkuk	6911 ujbel	6961 ujomo
6512 udfew	6562 udovi	6612 ufbod	6662 ufloh	6712 ufzel	6762 ugkan	6812 ugups	6862 uhlav	6912 ujbiv	6962 ujong
6513 udfod	6563 udoxa	6613 ufbum	6663 uflut	6713 ugafo	6763 ugkey	6813 uguxu	6863 uhlep	6913 ujboy	6963 ujose
6514 udfri	6564 udpac	6614 ufbyc	6664 uflyr	6714 ugahy	6764 ugkip	6814 uguza	6864 uhlir	6914 ujbud	6964 ujpad
6515 udfum	6565 udpeg	6615 ufcap	6665 ufmac	6715 ugaix	6765 ugkod	6815 ugvac	6865 uhlok	6915 ujcac	6965 ujpeo
6516 udfys	6566 udpin	6616 ufces	6666 ufmeg	6716 ugang	6766 ugkum	6816 ugveg	6866 uhlyc	6916 ujceg	6966 ujpol
6517 udgap	6567 udplu	6617 ufciv	6667 ufmif	6717 ugape	6767 uglap	6817 ugvom	6867 uhmat	6917 ujclu	6967 ujpub
6518 udges	6568 udpom	6618 ufcof	6668 ufmom	6718 ugara	6768 uglof	6818 ugwak	6868 uhmod	6918 ujcys	6968 ujraf
6519 udgof	6569 udpyx	6619 ufcuc	6669 ufnew	6719 ugaul	6769 ugluc	6819 ugwic	6869 uhmup	6919 ujdan	6969 ujred
6520 udguc	**6570** udrad	**6620** ufcyx	**6670** ufnik	**6720** ugbaf	**6770** ugmar	**6820** ugzax	**6870** uhnak	**6920** ujdey	**6970** ujril
6521 udgym	6571 udrec	6621 ufdar	6671 ufnol	6721 ugbed	6771 ugmez	6821 ugzer	6871 uhnec	6921 ujdif	6971 ujroc
6522 udhak	6572 udrol	6622 ufdez	6672 ufobo	6722 ugbit	6772 ugmid	6822 uhabs	6872 uhnut	6922 ujdod	6972 ujruz
6523 udhey	6573 udrub	6623 ufdip	6673 ufome	6723 ugboc	6773 ugmog	6823 uhajo	6873 uhoby	6923 ujdut	6973 ujsag
6524 udhip	6574 udsaf	6624 ufdog	6674 ufonc	6724 ugbry	6774 ugmux	6824 uhake	6874 uhocs	6924 ujego	6974 ujseb
6525 udhux	6575 udsed	6625 ufdux	6675 ufopa	6725 ugbuz	6775 ugmys	6825 uhalu	6875 uhogn	6925 ujeli	6975 ujsik
6526 udich	6576 udsby	6626 ufeat	6676 ufori	6726 ugcag	6776 ugnas	6826 uharc	6876 uhoro	6926 ujerx	6976 ujsox
6527 udido	6577 udsig	6627 ufedy	6677 ufots	6727 ugceb	6777 ugnet	6827 uhavi	6877 uhotz	6927 ujeve	6977 ujsun
6528 udine	6578 udsoc	6628 ufefo	6678 ufpaf	6728 ugcil	6778 ugnor	6828 uhbex	6878 uhoxi	6928 ujews	6978 ujtah
6529 udisi	6579 udsuz	6629 ufelv	6679 ufped	6729 ugcox	6779 ugnug	6829 uhcah	6879 uhpaj	6929 ujfap	6979 ujtim
6530 udiza	**6580** udtav	**6630** ufemu	**6680** ufpil	**6730** ugcun	**6780** ugnyx	**6830** uhcig	**6880** uhpho	**6930** ujfit	**6980** ujuco
6531 udjar	6581 udtem	6631 uferg	6681 ufpoc	6731 ugcyd	6781 ugoga	6831 uhcul	6881 uhpuy	6931 ujfro	6981 ujuda
6532 udjez	6582 udtis	6632 ufese	6682 ufpuz	6732 ugday	6782 ugolu	6832 uhdet	6882 uhric	6932 ujful	6982 ujuky
6533 udjog	6583 udtuf	6633 uffas	6683 ufrag	6733 ugdef	6783 ugopi	6833 uhdom	6883 uhroh	6933 ujgar	6983 ujume
6534 udkas	6584 udubs	6634 uffix	6684 ufreb	6734 ugdim	6784 ugovo	6834 uhdri	6884 uhsib	6934 ujgez	6984 ujuph
6535 udkek	6585 uduck	6635 uffor	6685 ufrid	6735 ugdoz	6785 ugoxe	6835 uhdyl	6885 uhsko	6935 ujgog	6985 ujuvi
6536 udkid	6586 uduly	6636 uffug	6686 ufrox	6736 ugduv	6786 ugpav	6836 uhebi	6886 uhsla	6936 ujhat	6986 ujvay
6537 udkor	6587 uduno	6637 ufgav	6687 ufrun	6737 ugect	6787 ugpex	6837 uheft	6887 uhswe	6937 ujhib	6987 ujvef
6538 udkug	6588 udura	6638 ufgim	6688 ufryf	6738 ugegu	6788 ugpli	6838 uhelk	6888 uhtan	6938 ujhov	6988 ujvid
6539 udkyp	6589 uduzi	6639 ufgop	6689 ufsaw	6739 ugeko	6789 ugpop	6839 uhena	6889 uhtop	6939 ujhuf	6989 ujvoz
6540 udlaw	**6590** udvag	**6640** ufgur	**6690** ufsef	**6740** ugems	**6790** ugpur	**6840** uherp	**6890** uhtux	**6940** ujhyg	**6990** ujwaw
6541 udleh	6591 udvil	6641 ufgyn	6691 ufsit	6741 ugepy	6791 ugpyk	6841 uhesh	6891 uhual	6941 ujiba	6991 ujwem
6542 udlit	6592 udvox	6642 ufhab	6692 ufsoz	6742 ugexa	6792 ugraz	6842 uhfol	6892 uhubu	6942 ujife	6992 ujwis
6543 udlur	6593 udvun	6643 ufhek	6693 ufsuv	6743 ugezi	6793 ugrep	6843 uhfra	6893 uhufi	6943 ujimi	6993 ukabo
6544 udlyd	6594 udwir	6644 ufhig	6694 ufsys	6744 ugfah	6794 ugrif	6844 uhfuf	6894 uhvis	6944 ujiru	6994 ukacs
6545 udmax	6595 udwob	6645 ufhon	6695 uften	6745 ugfem	6795 ugrok	6845 uhgaz	6895 uhvon	6945 ujiso	6995 ukald
6546 udmer	6596 udyet	6646 ufimo	6696 uftha	6746 ugfyl	6796 ugryc	6846 uhgli	6896 uhwaf	6946 ujkav	6996 ukapy
6547 udmib	6597 udzic	6647 ufiny	6697 uftry	6747 uggal	6797 ugsab	6847 uhgoc	6897 uhwel	6947 ujkew	6997 ukarm
6548 udmov	6598 udzuk	6648 ufipu	6698 uftuk	6748 uggen	6798 ugsev	6848 uhgre	6898 uhwip	6948 ujkiz	6998 ukate
6549 udmut	6599 ufabe	6649 ufiwa	6699 ufuba	6749 uggiv	6799 ugsis	6849 uhgud	6899 uhwog	6949 ujkry	6999 ukauk

ukawn ukaxi ukaza ukbac

SECTION B

7000 ukbeg	**7050** ukken	**7100** ukwet	**7150** ulkut	**7200** umbyx	**7250** ummik	**7300** unami	**7350** unlec	**7400** upaku	**7450** upono
7001 ukbif	7051 ukkos	7101 ukwod	7151 ullab	7201 umcaw	7251 ummop	7301 unanz	7351 unliv	7401 upbaz	7451 upork
7002 ukblu	7052 ukkyk	7102 ukyew	7152 ulley	7202 umcir	7252 ummur	7302 unapa	7352 unlol	7402 upbok	7452 upovu
7003 ukbro	7053 uklam	7103 ukyla	7153 ullop	7203 umcoy	7253 ummyd	7303 unask	7353 unlub	7403 upbre	7453 uppes
7004 ukcad	7054 uklel	7104 ukyus	7154 ulluk	7204 umcut	7254 umnax	7304 unath	7354 unmaf	7404 upbus	7454 uppiz
7005 ukcec	7055 uklon	7105 ukzah	7155 ulmew	7205 umdac	7255 umner	7305 unauc	7355 unmed	7405 upbyt	7455 uppro
7006 ukcol	7056 uklud	7106 ukzin	7156 ulmoy	7206 umdeg	7256 umnif	7306 unaxo	7356 unmoc	7406 upcel	7456 uppuc
7007 ukcri	7057 ukman	7107 ulaep	7157 ulnin	7207 umdin	7257 umnov	7307 unaye	7357 unmuz	7407 upcis	7457 uppyd
7008 ukcub	7058 ukmex	7108 ulait	7158 ulnos	7208 umdul	7258 umnup	7308 unbez	7358 unnag	7408 upcyp	7458 uprit
7009 ukcyn	7059 ukmig	7109 ulary	7159 uloaf	7209 umdyp	7259 umnyg	7309 unbly	7359 unneb	7409 upduy	7459 upryx
7010 ukdaf	**7060** ukmoh	**7110** uland	**7160** uloen	**7210** umefi	**7260** umoch	**7310** unbog	**7360** unnox	**7410** upecy	**7460** upsas
7011 ukded	7061 ukmum	7111 ulbat	7161 uloid	7211 umega	7261 umode	7311 unbux	7361 unobi	7411 upeka	7461 upsek
7012 ukdil	7062 ukmyl	7112 ulbeb	7162 ulora	7212 umeke	7262 umoja	7312 uncin	7362 unodu	7412 upelb	7462 upsor
7013 ukdoc	7063 uknap	7113 ulbow	7163 uloux	7213 umelm	7263 umomi	7313 undav	7363 unoft	7413 upenu	7463 upspi
7014 ukduz	7064 uknes	7114 ulbul	7164 ulows	7214 umepu	7264 umops	7314 undej	7364 unogo	7414 upfec	7464 uptac
7015 ukdyx	7065 uknot	7115 ulcal	7165 ulozo	7215 umery	7265 umorf	7315 undib	7365 unoha	7415 upflo	7465 uptil
7016 ukeba	7066 ukobu	7116 ulcef	7166 ulpip	7216 umeth	7266 umotu	7316 undop	7366 unove	7416 upfub	7466 uptof
7017 ukech	7067 ukogy	7117 ulcim	7167 ulpla	7217 umfam	7267 umpaz	7317 unebs	7367 unoys	7417 upgir	7467 upugo
7018 ukedo	7068 ukols	7118 ulcot	7168 ulpof	7218 umfel	7268 umpep	7318 unegi	7368 unpaw	7418 upgle	7468 upuli
7019 ukege	7069 ukond	7119 uldad	7169 ulrah	7219 umfig	7269 umpit	7319 unema	7369 unpir	7419 uphah	7469 upush
7020 ukeki	**7070** ukorc	**7120** uldec	**7170** ulrog	**7220** umfot	**7270** umpok	**7320** uneno	**7370** unpoz	**7420** upheg	**7470** uputa
7021 ukemp	7071 ukoto	7121 uldol	7171 ulrys	7221 umfud	7271 umpus	7321 unepe	7371 unrak	7421 uphol	7471 upvav
7022 ukent	7072 ukoze	7122 uldub	7172 ulsak	7222 umgan	7272 umrab	7322 unert	7372 unrem	7422 uphur	7472 upvet
7023 ukesu	7073 ukpar	7123 uldym	7173 ulshu	7223 umgew	7273 umrev	7323 unezu	7373 unrob	7423 upiby	7473 upvif
7024 ukfag	7074 ukpez	7124 ulemb	7174 ulsme	7224 umgic	7274 umriz	7324 unfax	7374 unruf	7424 upico	7474 upvoy
7025 ukfeb	7075 ukpik	7125 uleni	7175 ultag	7225 umgod	7275 umron	7325 unfer	7375 unsal	7425 upifa	7475 upwom
7026 ukfir	7076 ukpog	7126 uleps	7176 ultek	7226 umgum	7276 umruk	7326 ungay	7376 unscu	7426 upils	7476 upyca
7027 ukfox	7077 ukpyt	7127 ulerz	7177 ultho	7227 umhet	7277 umsad	7327 ungep	7377 unsen	7427 upitu	7477 upyem
7028 ukfun	7078 ukras	7128 uletu	7178 ultuz	7228 umhor	7278 umsec	7328 ungil	7378 unski	7428 upive	7478 upzaf
7029 ukgat	7079 ukrek	7129 ulfed	7179 ulupu	7229 umici	7279 umsol	7329 ungus	7379 unsos	7429 upizi	7479 upzib
7030 ukgef	**7080** ukror	**7130** ulfis	**7180** ulusi	**7230** umigy	**7280** umspy	**7330** unhaz	**7380** untap	**7430** upjef	**7480** upzly
7031 ukgid	7081 ukrug	7131 ulgaw	7181 ulvur	7231 umiku	7281 umsti	7331 unhex	7381 untew	7431 upjid	7481 upzoc
7032 ukgoz	7082 uksav	7132 ulget	7182 ulwer	7232 umilo	7282 umsub	7332 unhik	7382 untiz	7432 upjot	7482 upzuz
7033 ukgul	7083 uksim	7133 ulgig	7183 ulwiz	7233 umima	7283 umted	7333 unhum	7383 untod	7433 upjup	7483 uqaha
7034 ukgyb	7084 uksop	7134 ulgro	7184 ulxem	7234 umise	7284 umtip	7334 unhyp	7384 untur	7434 upkak	7484 uqain
7035 ukhaw	7085 uksur	7135 ulguy	7185 ulxyn	7235 umjap	7285 umtox	7335 unide	7385 untyl	7435 upkeh	7485 uqalp
7036 ukhut	7086 uktab	7136 ulhan	7186 ulyap	7236 umjes	7286 umtuc	7336 unili	7386 unuca	7436 upkob	7486 uqamb
7037 ukhyc	7087 ukthy	7137 ulhoc	7187 ulyel	7237 umjil	7287 umtyt	7337 unimu	7387 unudo	7437 upkys	7487 uqark
7038 ukibi	7088 uktit	7138 ulhyl	7188 umacy	7238 umjof	7288 umubo	7338 unina	7388 unump	7438 uplov	7488 uqbaw
7039 ukicu	7089 uktow	7139 uliha	7189 umadu	7239 umjux	7289 umuka	7339 unist	7389 ununt	7439 uplul	7489 uqboj
7040 ukida	**7090** uktre	**7140** ulims	**7190** umafa	**7240** umkar	**7290** umulu	**7340** unjab	**7390** unure	**7440** uplyg	**7490** uqcat
7041 ukiko	7091 ukube	7141 ulion	7191 umale	7241 umkib	7291 umune	7341 unjif	7391 unvam	7441 upmev	7491 uqcen
7042 uking	7092 ukuja	7142 ulirp	7192 umaph	7242 umkog	7292 umvat	7342 unjon	7392 unvel	7442 upmim	7492 uqclo
7043 ukisy	7093 ukult	7143 ulitz	7193 umarx	7243 umlas	7293 umvix	7343 unjuk	7393 unvij	7443 upmon	7493 uqcuf
7044 ukith	7094 ukumy	7144 uljav	7194 umato	7244 umlek	7294 umvoc	7344 unkac	7394 unvot	7444 upmuk	7494 uqdew
7045 ukjak	7095 ukunz	7145 uljib	7195 umbak	7245 umlow	7295 umwal	7345 unkeg	7395 unwan	7445 upnan	7495 uqdir
7046 ukjem	7096 ukver	7146 uljod	7196 umbem	7246 umlug	7296 umwiv	7346 unkit	7396 unyat	7446 upney	7496 uqdot
7047 ukjob	7097 ukvov	7147 uljum	7197 umbid	7247 umlyb	7297 umxis	7347 unkul	7397 unzeh	7447 upnum	7497 uqdum
7048 ukjuf	7098 ukvup	7148 ulkay	7198 umbob	7248 ummaj	7298 umyom	7348 unkyn	7398 unzof	7448 upogi	7498 uqeal
7049 ukkal	7099 ukwaz	7149 ulkex	7199 umbuf	7249 ummeh	7299 umzoz	7349 unlad	7399 upade	7449 upomy	7499 uqeco

7500 uqeig	**7550** uqwov	**7600** urkoz	**7650** urxur	**7700** ustak	**7750** uthyb	**7800** utudu	**7850** uvger	**7900** uvsip	**7950** uwfom
7501 uqers	7551 uqxan	7601 urkru	7651 uryaf	7701 ustij	7751 utiar	7801 utums	7851 uvgib	7901 uvsob	7951 uwfru
7502 uqfaf	7552 uqxor	7602 urlah	7652 urzad	7702 ustoc	7752 utied	7802 utunk	7852 uvgov	7902 uvsuf	7952 uwgad
7503 uqfli	7553 uracu	7603 urlen	7653 urzig	7703 ustyd	7753 utiky	7803 utuve	7853 uvgup	7903 uvtep	7953 uwgec
7504 uqfos	7554 urams	7604 urlos	7654 urzol	7704 usugi	7754 utilu	7804 utuxa	7854 uvhaj	7904 uvtiv	7954 uwgif
7505 uqfre	7555 uraro	7605 urluw	7655 usafy	7705 usunu	7755 utins	7805 utvex	7855 uvhir	7905 uvtul	7955 uwgol
7506 uqham	7556 urata	7606 urmam	7656 usals	7706 usupo	7756 utipa	7806 utvig	7856 uvhox	7906 uvudy	7956 uwgub
7507 uqheb	7557 urave	7607 urmel	7657 usamu	7707 usurb	7757 utixi	7807 utvyn	7857 uvhuk	7907 uvval	7957 uwhav
7508 uqhog	7558 uraxy	7608 urmir	7658 usank	7708 usute	7758 utjaz	7808 utwen	7858 uvhys	7908 uvven	7958 uwhit
7509 uqhud	7559 urazi	7609 urmot	7659 usapi	7709 usveh	7759 utjec	7809 utwil	7859 uvics	7909 uvvit	7959 uwhoy
7510 uqiap	**7560** urbab	**7610** urmud	**7660** usavo	**7710** usvip	**7760** utjin	**7810** utxas	**7860** uvigo	**7910** uvvos	**7960** uwhul
7511 uqile	7561 urbev	7611 urnar	7661 usawa	7711 usvus	7761 utjol	7811 utxer	7861 uvika	7911 uvway	7961 uwibe
7512 uqink	7562 urbon	7612 urnez	7662 usbaj	7712 usvyt	7762 utkik	7812 utyok	7862 uviph	7912 uvwex	7962 uwica
7513 uqkah	7563 urbuk	7613 urnip	7663 usbil	7713 uswav	7763 utkux	7813 utyuc	7863 uvizu	7913 uvxin	7963 uwifo
7514 uqket	7564 urcem	7614 urnog	7664 usbys	7714 uswec	7764 utlak	7814 utzep	7864 uvjat	7914 uvzan	7964 uwigu
7515 uqkow	7565 urciz	7615 urnux	7665 uscey	7715 uswon	7765 utlev	7815 uvahs	7865 uvjew	7915 uvzik	7965 uwimp
7516 uqkri	7566 urcob	7616 urnyk	7666 usdex	7716 usxol	7766 utlor	7816 uvaki	7866 uvkab	7916 uvzum	7966 uwits
7517 uqlaz	7567 urcra	7617 uroco	7667 usejo	7717 usyba	7767 utlys	7817 uvalb	7867 uvkev	7917 uwaba	7967 uwjaf
7518 uqlic	7568 urcyc	7618 urohm	7668 usene	7718 uszid	7768 utmaw	7818 uvamt	7868 uvkis	7918 uwach	7968 uwjed
7519 uqlus	7569 urdap	7619 uroki	7669 usfik	7719 uszle	7769 utmyx	7819 uvawe	7869 uvklu	7919 uwaim	7969 uwjoc
7520 uqmay	**7570** urdof	**7620** urons	**7670** usgno	**7720** utahi	**7770** utnad	**7820** uvaya	**7870** uvkon	**7920** uwaju	**7970** uwkag
7521 uqmis	7571 urduc	7621 urpan	7671 usgry	7721 utaka	7771 utnir	7821 uvbap	7871 uvkyr	7921 uwaly	7971 uwkeb
7522 uqmul	7572 ureha	7622 urphy	7672 ushax	7722 utalo	7772 utnul	7822 uvbes	7872 uvlac	7922 uwant	7972 uwkil
7523 uqnex	7573 urelu	7623 urpib	7673 usidy	7723 utamp	7773 utony	7823 uvbig	7873 uvleg	7923 uwase	7973 uwkun
7524 uqnil	7574 ureny	7624 urpod	7674 usifi	7724 utanu	7774 utoph	7824 uvbof	7874 uviif	7924 uwazo	7974 uwkyd
7525 uqnub	7575 urepo	7625 urpum	7675 usita	7725 utard	7775 utore	7825 uvbuc	7875 uvlom	7925 uwbax	7975 uwlay
7526 uqofa	7576 urets	7626 urrac	7676 usjul	7726 utasy	7776 utout	7826 uvbyb	7876 uvlyp	7926 uwber	7976 uwlef
7527 uqoge	7577 urexi	7627 urreg	7677 usker	7727 utaxt	7777 utoso	7827 uvcar	7877 uvmad	7927 uwblo	7977 uwlig
7528 uqolm	7578 ureze	7628 urrom	7678 uskig	7728 utbah	7778 utova	7828 uvchy	7878 uvmec	7928 uwbup	7978 uwloz
7529 uqopo	7579 urfay	7629 urrus	7679 uskup	7729 utbey	7779 utozi	7829 uvcic	7879 uvmix	7929 uwbyn	7979 uwluv
7530 uqorn	**7580** urfex	**7630** ursat	**7680** uskyf	**7730** utbot	**7780** utpew	**7830** uvcog	**7880** uvmol	**7930** uwcaz	**7980** uwmak
7531 uqosk	7581 urfil	7631 ursoy	7681 uslem	7731 utcid	7781 utpha	7831 uvcux	7881 uvnaf	7931 uwcep	7981 uwmem
7532 uqpru	7582 urfoh	7632 ursul	7682 usluf	7732 utdat	7782 utpis	7832 uvdas	7882 uvned	7932 uwcro	7982 uwmip
7533 uqser	7583 urfuz	7633 urtax	7683 uslyx	7733 utdru	7783 utple	7833 uvdek	7883 uvnoc	7933 uwcus	7983 uwmob
7534 uqsid	7584 urgas	7634 urter	7684 usmyr	7734 utean	7784 utpoy	7834 uvdor	7884 uvnuz	7934 uwdab	7984 uwmuf
7535 uqspa	7585 urgek	7635 urtid	7685 usnel	7735 utefu	7785 utram	7835 uvdug	7885 uvoca	7935 uwdev	7985 uwmyg
7536 uqsto	7586 urgit	7636 urtup	7686 usnud	7736 utely	7786 utrel	7836 uvdyc	7886 uvodi	7936 uwdic	7986 uwnal
7537 uqtel	7587 urgor	7637 urtys	7687 usnyc	7737 uterk	7787 utroj	7837 uveid	7887 uvoho	7937 uwdon	7987 uwnek
7538 uqtif	7588 urgug	7638 uruby	7688 usocu	7738 utevo	7788 utruw	7838 uvelz	7888 uvonu	7938 uwdry	7988 uwnix
7539 uqtom	7589 urhec	7639 urufu	7689 usoko	7739 utfal	7789 utsax	7839 uvemy	7889 uvost	7939 uwduk	7989 uwobs
7540 uqtug	**7590** urhiv	**7640** urulk	**7690** usoma	**7740** utfet	**7790** utske	**7840** uvept	**7890** uvoxy	**7940** uwebu	**7990** uwoda
7541 uqula	7591 urids	7641 urvaz	7691 usope	7741 utfif	7791 utsov	7841 uvere	7891 uvpag	7941 uwefa	7991 uwofi
7542 uqumo	7592 uriga	7642 urvey	7692 usosy	7742 utfoc	7792 utspu	7842 uveso	7892 uvpeb	7942 uwele	7992 uwojo
7543 uqupi	7593 urime	7643 urvim	7693 usres	7743 utgeh	7793 utsyc	7843 uvevu	7893 uvpim	7943 uwens	7993 uwoku
7544 uquvu	7594 urinu	7644 urvok	7694 usriv	7744 utgow	7794 uttaf	7844 uvfav	7894 uvpow	7944 uweph	7994 uwowl
7545 uqvep	7595 uripy	7645 urvuf	7695 usrof	7745 utgra	7795 uttes	7845 uvfey	7895 uvpun	7945 uwerd	7995 uwpam
7546 uqvod	7596 uriri	7646 urwin	7696 usruc	7746 uthag	7796 uttic	7846 uvfiz	7896 uvraw	7946 uweto	7996 uwpel
7547 uqwad	7597 urixo	7647 urwow	7697 usryb	7747 uthiz	7797 uttog	7847 uvfop	7897 uvref	7947 uwfac	7997 uwpiv
7548 nqweg	7598 urjaw	7648 urwry	7698 ussuk	7748 uthop	7798 uttym	7848 uvfur	7898 uvsah	7948 uwfeg	7998 uwpot
7549 uqwim	7599 urkef	7649 urxal	7699 ussyn	7749 uthus	7799 utuch	7849 uvgax	7899 uvsem	7949 uwfib	7999 uwpud

uwrew uwrik uwrod uwryl

[redacted]

SECTION B

8000 uwsap	**8050** uxfoy	**8100** uxrup	**8150** uzemi	**8200** uzown	**8250** vaful	**8300** vavay	**8350** veilk	**8400** veuso	**8450** vipip
8001 uwses	8051 uxfut	8101 uxryd	8151 uzest	8201 uzoxo	8251 vagez	8301 vavef	8351 veite	8401 vevad	8451 vipla
8002 uwsiz	8052 uxgam	8102 uxsaz	8152 uzewe	8202 uzpak	8252 vahib	8302 vavoz	8352 vejan	8402 vevec	8452 vipof
8003 uwsof	8053 uxgel	8103 uxsep	8153 uzfaz	8203 uzpid	8253 vahov	8303 vawaw	8353 vejik	8403 vevol	8453 vipre
8004 uwsuc	8054 uxgip	8104 uxsmi	8154 uzfep	8204 uzplo	8254 vahyg	8304 vazal	8354 vekap	8404 vewac	8454 virah
8005 uwsyr	8055 uxgot	8105 uxsow	8155 uzfin	8205 uzpry	8255 vaiba	8305 vazen	8355 vekes	8405 veyls	8455 virog
8006 uwtaw	8056 uxgru	8106 uxsty	8156 uzfly	8206 uzpuf	8256 vaife	8306 vazos	8356 vekim	8406 veyoc	8456 visak
8007 uwtex	8057 uxher	8107 uxsus	8157 uzfok	8207 uzral	8257 vaimi	8307 vazug	8357 vekuc	8407 vibat	8457 vitek
8008 uwthi	8058 uxhis	8108 uxtal	8158 uzfus	8208 uzren	8258 vairu	8308 veaby	8358 velid	8408 vibeb	8458 vitho
8009 uwtor	8059 uxhod	8109 uxteg	8159 uzgab	8209 uzros	8259 vaiso	8309 veadz	8359 velog	8409 vibir	8459 vituz
8010 uwtuy	**8060** uxidi	**8110** uxtin	**8160** uzgev	**8210** uzrut	**8260** vajas	**8310** vealc	**8360** velux	**8410** vibow	**8460** viupu
8011 uwuga	8061 uxifu	8111 uxtos	8161 uzgis	8211 uzsam	8261 vajor	8311 veana	8361 velym	8411 vibul	8461 viusi
8012 uwumi	8062 uxilk	8112 uxtum	8162 uzgon	8212 uzsel	8262 vakav	8312 veari	8362 vemas	8412 vicef	8462 vivur
8013 uwurn	8063 uxisa	8113 uxtyx	8163 uzgyr	8213 uzsho	8263 vakew	8313 veasm	8363 vemek	8413 vidad	8463 viwam
8014 uwutu	8064 uxite	8114 uxuki	8164 uzhap	8214 uzsif	8264 vakiz	8314 veaxu	8364 vemor	8414 videc	8464 viwer
8015 uwvar	8065 uxjan	8115 uxung	8165 uzhew	8215 uzsud	8265 vakry	8315 vebag	8365 vemug	8415 vidol	8465 viwiz
8016 uwvez	8066 uxjeb	8116 uxupa	8166 uzhof	8216 uzsyk	8266 valer	8316 vebef	8366 venav	8416 vidub	8466 vixem
8017 uwvin	8067 uxjik	8117 uxuso	8167 uzhuc	8217 uztat	8267 vamaz	8317 vebib	8367 veneh	8417 vieda	8467 vixyn
8018 uwvog	8068 uxkap	8118 uxvad	8168 uzhyd	8218 uztez	8268 vamin	8318 vebox	8368 venop	8418 viemb	8468 vizez
8019 uxaby	8069 uxkes	8119 uxvec	8169 uzibs	8219 uztig	8269 vamok	8319 vebun	8369 venur	8419 vieps	8469 vizuc
8020 uxadz	**8070** uxkim	**8120** uxvir	**8170** uzice	**8220** uztoy	**8270** vanev	**8320** vebyl	**8370** veobe	**8420** vietu	**8470** voabi
8021 uxaho	8071 uxkuc	8121 uxvol	8171 uzipi	8221 uzufo	8271 vanig	8321 vecay	8371 veock	8421 viexo	8471 voact
8022 uxalc	8072 uxlar	8122 uxwac	8172 uzirl	8222 uzuma	8272 vanuk	8322 vecet	8372 veolo	8422 vifed	8472 voady
8023 uxana	8073 uxlez	8123 uxwed	8173 uzkad	8223 uzund	8273 vanyp	8323 vecoz	8373 veomu	8423 vifis	8473 voage
8024 uxari	8074 uxlid	8124 uxzat	8174 uzkec	8224 uzuse	8274 vaody	8324 vedak	8374 veoni	8424 vigaw	8474 voama
8025 uxasm	8075 uxlog	8125 uxzom	8175 uzkir	8225 uzuti	8275 vaola	8325 vedem	8375 veopy	8425 viget	8475 voapo
8026 uxbag	8076 uxlym	8126 uzabu	8176 uzkol	8226 uzvan	8276 vaomo	8326 vedig	8376 veota	8426 vigig	8476 vobav
8027 uxbef	8077 uxmas	8127 uzaca	8177 uzkub	8227 uzvik	8277 vaong	8327 vedob	8377 vepab	8427 vigro	8477 vobij
8028 uxbib	8078 uxmek	8128 uzads	8178 uzlaf	8228 uzvum	8278 vaose	8328 veduf	8378 vepev	8428 viguy	8478 voble
8029 uxbox	8079 uxmit	8129 uzafi	8179 uzled	8229 uzwar	8279 vapec	8329 veebo	8379 vepif	8429 vihan	8479 voboh
8030 uxbun	**8080** uxmor	**8130** uzago	**8180** uzlip	**8230** uzwes	**8280** vapij	**8330** veede	**8380** vepon	**8430** vihoc	**8480** vobur
8031 uxbyl	8081 uxmug	8131 uzalm	8181 uzloc	8231 uzwot	8281 vapol	8331 veefy	8381 vepuk	8431 vihyl	8481 vocax
8032 uxcay	8082 uxnav	8132 uzane	8182 uzluz	8232 vaaci	8282 vapub	8332 veeku	8382 vepyc	8432 vijav	8482 vocer
8033 uxcet	8083 uxneh	8133 uzaty	8183 uzlyn	8233 vaafu	8283 vapyr	8333 veela	8383 verov	8433 vijib	8483 vocib
8034 uxcoz	8084 uxnop	8134 uzawk	8184 uzmag	8234 vabel	8284 varaf	8334 veern	8384 veryd	8434 vijod	8484 vocov
8035 uxdak	8085 uxnur	8135 uzbas	8185 uzmeb	8235 vabiv	8285 varil	8335 veeti	8385 vesaz	8435 vijum	8485 vocup
8036 uxdem	8086 uxobe	8136 uzbek	8186 uzmil	8236 vacac	8286 varuz	8336 vefaw	8386 vesep	8436 vikay	8486 vodaz
8037 uxdig	8087 uxock	8137 uzbli	8187 uzmox	8237 vaceg	8287 vaseb	8337 vefen	8387 vesmi	8437 vikex	8487 vodep
8038 uxdob	8088 uxolo	8138 uzbor	8188 uzmun	8238 vaclu	8288 vasik	8338 vefic	8388 vesow	8438 viklo	8488 vodix
8039 uxduf	8089 uxomu	8139 uzbug	8189 uzmyc	8239 vacom	8289 vasyd	8339 vefoy	8389 vesty	8439 vikut	8489 vodok
8040 uxdyn	**8090** uxoni	**8140** uzcav	**8190** uznah	**8240** vacys	**8290** vatah	**8340** vefut	**8390** vesus	**8440** vilab	**8490** vodus
8041 uxede	8091 uxopy	8141 uzcit	8191 uznef	8241 vadan	8291 vatob	8341 vegam	8391 vetal	8441 vilop	8491 voedi
8042 uxefy	8092 uxota	8142 uzcop	8192 uznib	8242 vadey	8292 vatyn	8342 vegel	8392 veteg	8442 viluk	8492 voemo
8043 uxeku	8093 uxpab	8143 uzcur	8193 uznoz	8243 vadif	8293 vauco	8343 vegip	8393 vetin	8443 vimew	8493 voeng
8044 uxela	8094 uxpev	8144 uzcyg	8194 uznyl	8244 vadod	8294 vauda	8344 vegot	8394 vetos	8444 vimoy	8494 voerl
8045 uxern	8095 uxpif	8145 uzdax	8195 uzoba	8245 vadut	8295 vauky	8345 vegru	8395 vetyx	8445 vinin	8495 voesy
8046 uxeti	8096 uxpon	8146 uzder	8196 uzoci	8246 vaeja	8296 vaume	8346 vehaf	8396 veucu	8446 vinyt	8496 voewa
8047 uxfaw	8097 uxpuk	8147 uzdiz	8197 uzoky	8247 vaeli	8297 vauph	8347 veher	8397 veuki	8447 viora	8497 voexe
8048 uxfen	8098 uxpyc	8148 uzdov	8198 uzord	8248 vaerx	8298 vauvi	8348 vehuz	8398 veung	8448 viows	8498 vofab
8049 uxfic	8099 uxrov	8149 uzdup	8199 uzote	8249 vafes	8299 vauzu	8349 veifu	8399 veupa	8449 viozo	8499 vofev

8500 vofip	8550 vorat	8600 vudup	8650 vuoba	8700 vyfaf	8750 wabow	8800 waows	8850 wefez	8900 weryb	8950 widuv
8501 voflu	8551 vorez	8601 vuemi	8651 vuoci	8701 vyfli	8751 wabul	8801 waozo	8851 wefog	8901 wesac	8951 wiect
8502 vofon	8552 vorhi	8602 vuepa	8652 vuoky	8702 vyfos	8752 wacal	8802 wapla	8852 wefux	8902 weseg	8952 wiems
8503 vogac	8553 vorow	8603 vuero	8653 vuord	8703 vyfre	8753 wacef	8803 wapip	8853 wegaf	8903 wesir	8953 wiepy
8504 vogeg	8554 vosew	8604 vuest	8654 vuote	8704 vyglu	8754 wacim	8804 wapof	8854 weged	8904 wesom	8954 wiexa
8505 vogin	8555 vosil	8605 vuewe	8655 vuown	8705 vyham	8755 wacot	8805 wapre	8855 wegno	8905 wesuk	8955 wiezi
8506 vogom	8556 vosod	8606 vufaz	8656 vuoxo	8706 vyheb	8756 wadad	8806 warah	8856 wegry	8906 wesyn	8956 wifah
8507 vogyl	8557 vosum	8607 vufep	8657 vupak	8707 vyhog	8757 wadec	8807 warog	8857 weguz	8907 wetak	8957 wifem
8508 vohad	8558 votas	8608 vufin	8658 vupem	8708 vyhud	8758 wadol	8808 wasak	8858 wehax	8908 wetet	8958 wifyl
8509 vohot	8559 votey	8609 vufly	8659 vupid	8709 vyibo	8759 wadub	8809 washu	8859 wehub	8909 wetij	8959 wigal
8510 vohug	8560 votik	8610 vufok	8660 vuplo	8710 vykah	8760 wadym	8810 watag	8860 weick	8910 wetoc	8960 wigen
8511 vohyr	8561 votol	8611 vufus	8661 vupry	8711 vyket	8761 waeda	8811 watek	8861 weidy	8911 wetun	8961 wigiv
8512 voicy	8562 votut	8612 vugab	8662 vupuf	8712 vykow	8762 waemb	8812 watho	8862 weifi	8912 wetyd	8962 wigut
8513 voigi	8563 voufa	8613 vugev	8663 vural	8713 vylaz	8763 waeni	8813 watuz	8863 weips	8913 weugi	8963 wiheh
8514 voinz	8564 vounc	8614 vugis	8664 vuros	8714 vylic	8764 waeps	8814 waupu	8864 weire	8914 weunu	8964 wihin
8515 voiro	8565 vourg	8615 vugon	8665 vurut	8715 vylus	8765 waerz	8815 wausi	8865 weita	8915 weupo	8965 wihuy
8516 voism	8566 voust	8616 vugyr	8666 vusam	8716 vymis	8766 waetu	8816 wavur	8866 wejul	8916 weurb	8966 wijel
8517 vojeh	8567 vouxo	8617 vuhap	8667 vusel	8717 vymul	8767 waexo	8817 wawam	8867 weker	8917 weusa	8967 wijig
8518 vojim	8568 vouze	8618 vuhew	8668 vusho	8718 vynex	8768 waeye	8818 wawer	8868 wekig	8918 weute	8968 wijus
8519 vojoy	8569 vovap	8619 vuhof	8669 vusif	8719 vynil	8769 wafed	8819 wawiz	8869 wekup	8919 weveh	8969 wikan
8520 vojub	8570 voves	8620 vuhuc	8670 vusud	8720 vynub	8770 wafis	8820 waxem	8870 wekyf	8920 wevip	8970 wikey
8521 vokaf	8571 voviv	8621 vuhyd	8671 vusyk	8721 vyofa	8771 wagaw	8821 waxyn	8871 welem	8921 wevow	8971 wikip
8522 voked	8572 vowor	8622 vuibs	8672 vutat	8722 vyoge	8772 wagig	8822 wayap	8872 welin	8922 wevus	8972 wikod
8523 vokli	8573 voxel	8623 vuice	8673 vutez	8723 vyolm	8773 wagro	8823 wayel	8873 welob	8923 wewav	8973 wikum
8524 vokoc	8574 voyar	8624 vuipi	8674 vutig	8724 vyopo	8774 waguy	8824 wazas	8874 weluf	8924 wewec	8974 wilap
8525 volag	8575 voyex	8625 vuirl	8675 vutru	8725 vyorn	8775 wahan	8825 wazez	8875 welyx	8925 wewon	8975 wiles
8526 voleb	8576 voyog	8626 vujac	8676 vuvan	8726 vyosk	8776 wahoc	8826 wazor	8876 wemal	8926 wexol	8976 wilof
8527 voliz	8577 vozek	8627 vujeg	8677 vuvik	8727 vyozu	8777 wahyl	8827 wazuc	8877 wemos	8927 weyba	8977 wiluc
8528 volox	8578 vozir	8628 vujuw	8678 vuvum	8728 vypef	8778 waiha	8828 weafy	8878 wemyr	8928 wezid	8978 wimar
8529 volun	8579 vuabu	8629 vukad	8679 vuwar	8729 vypru	8779 waims	8829 weamu	8879 wenam	8929 wezle	8979 wimez
8530 volyt	8580 vuaca	8630 vukec	8680 vuwes	8730 vyrix	8780 wairp	8830 weank	8880 wenel	8930 wezot	8980 wimid
8531 vomah	8581 vuads	8631 vukir	8681 vuwot	8731 vyser	8781 waitz	8831 weapi	8881 wenit	8931 wiafo	8981 wimog
8532 vomef	8582 vuafi	8632 vukol	8682 vuyaw	8732 vysid	8782 wajav	8832 weart	8882 wenud	8932 wiang	8982 wimux
8533 vomoz	8583 vuago	8633 vukub	8683 vuzey	8733 vyspa	8783 wajib	8833 weavo	8883 wenyc	8933 wiape	8983 wimys
8534 vonay	8584 vualm	8634 vulaf	8684 vuzog	8734 vysto	8784 wajod	8834 weawa	8884 weocu	8934 wibaf	8984 winas
8535 vonem	8585 vuane	8635 vuled	8685 vuzul	8735 vytar	8785 wajum	8835 weaxe	8885 weoko	8935 wibed	8985 winet
8536 vonid	8586 vuaty	8636 vulip	8686 vyaha	8736 vytel	8786 wakay	8836 webaj	8886 weoma	8936 wibit	8986 winor
8537 vonob	8587 vuawk	8637 vuloc	8687 vyain	8737 vytif	8787 wakex	8837 webil	8887 weope	8937 wiboc	8987 winug
8538 vonuf	8588 vubas	8638 vuluz	8688 vyalp	8738 vytom	8788 waklo	8838 weboz	8888 weosy	8938 wibry	8988 winyx
8539 vooak	8589 vubek	8639 vulyn	8689 vyamb	8739 vytug	8789 wakut	8839 wecod	8889 weoti	8939 wibuz	8989 wioga
8540 voofu	8590 vubli	8640 vumag	8690 vyark	8740 vyula	8790 walab	8840 wecum	8890 wepas	8940 wicag	8990 wiolu
8541 vooly	8591 vubor	8641 vumeb	8691 vybaw	8741 vyumo	8791 waley	8841 wedaw	8891 wepek	8941 wiceb	8991 wiovo
8542 voomp	8592 vubug	8642 vumil	8692 vyboj	8742 vyweg	8792 walil	8842 wedex	8892 wepix	8942 wicil	8992 wioxe
8543 voone	8593 vucit	8643 vumox	8693 vycat	8743 vywim	8793 walop	8843 wedis	8893 wepor	8943 wicox	8993 wipav
8544 voosi	8594 vucop	8644 vumun	8694 vyclo	8744 vywov	8794 waluk	8844 wedly	8894 wepug	8944 wicun	8994 wipex
8545 voowo	8595 vucyg	8645 vumyc	8695 vydew	8745 waand	8795 wamew	8845 wedur	8895 werap	8945 wicyd	8995 wipli
8546 vopal	8596 vudax	8646 vunah	8696 vydir	8746 waary	8796 wamoy	8846 weelf	8896 weres	8946 widay	8996 wipop
8547 vopen	8597 vuder	8647 vunef	8697 vydot	8747 wabat	8797 wanin	8847 weene	8897 weriv	8947 widef	8997 wipur
8548 vopig	8598 vudiz	8648 vunib	8698 vydum	8748 wabeb	8798 wanyt	8848 weeva	8898 werof	8948 widim	8998 wipyk
8549 vopos	8599 vudov	8649 vunoz	8699 vyers	8749 wabir	8799 waora	8849 weexu	8899 weruc	8949 widoz	8999 wiraz

wirep wirif wirok wirud

[redacted]

[redacted]

SECTION B

9000 wiryc	**9050** wokuk	**9100** wucat	**9150** wupef	**9200** wydig	**9250** wyomu	**9300** xafal	**9350** xaske	**9400** xeers	**9450** xetif
9001 wisev	9051 wolav	9101 wucen	9151 wuply	9201 wydob	9251 wyoni	9301 xafet	9351 xasli	9401 xefaf	9451 xetom
9002 wisis	9052 wolep	9102 wuclo	9152 wupru	9202 wyduf	9252 wyota	9302 xafif	9352 xasov	9402 xefli	9452 xetug
9003 wison	9053 wolir	9103 wucuf	9153 wuraj	9203 wyebo	9253 wypab	9303 xafoc	9353 xaspu	9403 xefos	9453 xetyb
9004 witec	9054 wolyc	9104 wudew	9154 wurey	9204 wyede	9254 wypev	9304 xageh	9354 xasyc	9404 xefre	9454 xeufy
9005 witub	9055 womey	9105 wudir	9155 wuser	9205 wyefy	9255 wypif	9305 xagow	9355 xataf	9405 xeglu	9455 xeula
9006 wityp	9056 womod	9106 wudot	9156 wusid	9206 wyeku	9256 wypon	9306 xagra	9356 xates	9406 xeham	9456 xeumo
9007 wiubi	9057 womup	9107 wudum	9157 wusky	9207 wyela	9257 wypuk	9307 xahag	9357 xatic	9407 xehog	9457 xeupi
9008 wiucy	9058 wonak	9108 wudys	9158 wuspa	9208 wyern	9258 wyrax	9308 xahem	9358 xatog	9408 xehud	9458 xeuvu
9009 wiuge	9059 wonec	9109 wueby	9159 wusto	9209 wyfaw	9259 wyrov	9309 xahiz	9359 xatym	9409 xeibo	9459 xevep
9010 wiulo	**9060** woniz	**9110** wueco	**9160** wusuh	**9210** wyfen	**9260** wyrup	**9310** xahop	**9360** xauch	**9410** xeile	**9460** xevod
9011 wiups	9061 wonof	9111 wuers	9161 wutel	9211 wyfic	9261 wysaz	9311 xahus	9361 xaudu	9411 xeink	9461 xewad
9012 wiuza	9062 wooby	9112 wufaf	9162 wutif	9212 wyfoy	9262 wysep	9312 xahyb	9362 xaums	9412 xejok	9462 xeweg
9013 wivac	9063 woocs	9113 wufli	9163 wutom	9213 wyfut	9263 wysmi	9313 xaiky	9363 xaunk	9413 xejun	9463 xewhy
9014 wiveg	9064 woogn	9114 wufos	9164 wutug	9214 wygam	9264 wysow	9314 xailu	9364 xauri	9414 xekah	9464 xewov
9015 wivom	9065 wooka	9115 wufre	9165 wutyb	9215 wygel	9265 wytal	9315 xains	9365 xauto	9415 xeket	9465 xeyac
9016 wiwak	9066 wooro	9116 wuglu	9166 wuvep	9216 wygip	9266 wyteg	9316 xaipa	9366 xauve	9416 xekow	9466 xezak
9017 wiwic	9067 wootz	9117 wuham	9167 wuvod	9217 wygot	9267 wytin	9317 xaixi	9367 xauxa	9417 xekri	9467 xezem
9018 wizax	9068 wooxi	9118 wuheb	9168 wuwad	9218 wygru	9268 wytum	9318 xajaz	9368 xavex	9418 xekyx	9468 xezob
9019 wizer	9069 wopaj	9119 wuhog	9169 wuweg	9219 wyhaf	9269 wyucu	9319 xajec	9369 xavig	9419 xelaz	9469 xezup
9020 woajo	**9070** wopho	**9120** wuhud	**9170** wuwhy	**9220** wyher	**9270** wyuki	**9320** xajin	**9370** xavyn	**9420** xelic	**9470** xiami
9021 woalu	9071 wopuy	9121 wuibo	9171 wuwim	9221 wyhis	9271 wyung	9321 xajol	9371 xawab	9421 xelus	9471 xianz
9022 woavi	9072 woric	9122 wuile	9172 wuwov	9222 wyhod	9272 wyupa	9322 xajud	9372 xawen	9422 xelyl	9472 xiapa
9023 wobex	9073 woroh	9123 wuink	9173 wuxan	9223 wyhuz	9273 wyuso	9323 xakik	9373 xawil	9423 xemay	9473 xiask
9024 wocah	9074 wosko	9124 wuixu	9174 wuxor	9224 wyidi	9274 wyvad	9324 xakof	9374 xawox	9424 xemis	9474 xiath
9025 wocig	9075 wostu	9125 wujip	9175 wuyac	9225 wyifu	9275 wyvec	9325 xakux	9375 xayez	9425 xemul	9475 xiaxo
9026 wocul	9076 woswe	9126 wujok	9176 wuyol	9226 wyilk	9276 wyvir	9326 xalak	9376 xayok	9426 xenex	9476 xibar
9027 wocym	9077 wosyf	9127 wujun	9177 wuzak	9227 wyisa	9277 wyvol	9327 xalev	9377 xayuc	9427 xenil	9477 xibez
9028 wodet	9078 wotux	9128 wukah	9178 wuzem	9228 wyite	9278 wywac	9328 xalor	9378 xazac	9428 xenoy	9478 xibim
9029 wodri	9079 woubu	9129 wuket	9179 wuziz	9229 wykap	9279 wywed	9329 xalys	9379 xazep	9429 xenub	9479 xibly
9030 wodyl	**9080** woude	**9130** wukow	**9180** wuzob	**9230** wykes	**9280** xaahi	**9330** xamaw	**9380** xazim	**9430** xeofa	**9480** xibog
9031 woebi	9081 woufi	9131 wukri	9181 wuzup	9231 wykim	9281 xaaka	9331 xamyx	9381 xazon	9431 xeoge	9481 xibux
9032 woena	9082 wowaf	9132 wukyx	9182 wyadz	9232 wykuc	9282 xaalo	9332 xanad	9382 xazuf	9432 xeolm	9482 xicas
9033 woerp	9083 wowel	9133 wulaz	9183 wyaho	9233 wylar	9283 xaamp	9333 xanir	9383 xeaha	9433 xeopo	9483 xicor
9034 woesh	9084 wowip	9134 wulic	9184 wyalc	9234 wylez	9284 xaanu	9334 xanul	9384 xealp	9434 xeorn	9484 xicug
9035 wofol	9085 wowog	9135 wulus	9185 wyana	9235 wylid	9285 xaard	9335 xaony	9385 xeamb	9435 xeosk	9485 xicyb
9036 wofra	9086 woxar	9136 wulyl	9186 wyari	9236 wylog	9286 xaasy	9336 xaoph	9386 xeark	9436 xeozu	9486 xidav
9037 wofuf	9087 woyef	9137 wumay	9187 wyasm	9237 wylux	9287 xaaxt	9337 xaore	9387 xeayo	9437 xepef	9487 xidej
9038 wogli	9088 woyne	9138 wumis	9188 wyaxu	9238 wymas	9288 xabah	9338 xaoso	9388 xebaw	9438 xeply	9488 xidib
9039 wogoc	9089 woyum	9139 wunex	9189 wybag	9239 wymek	9289 xabey	9339 xaova	9389 xeboj	9439 xepru	9489 xidop
9040 wogud	**9090** wozap	**9140** wunil	**9190** wybef	**9240** wymit	**9290** xabot	**9340** xaozi	**9390** xecen	**9440** xeraj	**9490** xiebs
9041 wohac	9091 wozif	9141 wunoy	9191 wybib	9241 wymor	9291 xacaj	9341 xapew	9391 xecuf	9441 xerey	9491 xiegi
9042 woimy	9092 wozus	9142 wunub	9192 wybox	9242 wymug	9292 xacid	9342 xapha	9392 xecyr	9442 xerix	9492 xiema
9043 woini	9093 wuaha	9143 wuofa	9193 wybun	9243 wynav	9293 xacos	9343 xapis	9393 xedew	9443 xeser	9493 xieno
9044 woisk	9094 wualp	9144 wuoge	9194 wybyl	9244 wyneh	9294 xadat	9344 xaple	9394 xedir	9444 xesid	9494 xiepe
9045 woizo	9095 wuamb	9145 wuolm	9195 wycay	9245 wynop	9295 xadru	9345 xapoy	9395 xedot	9445 xesky	9495 xiert
9046 wojad	9096 wuark	9146 wuopo	9196 wycet	9246 wynur	9296 xaefu	9346 xaram	9396 xedum	9446 xespa	9496 xifax
9047 wojek	9097 wuayo	9147 wuorn	9197 wycoz	9247 wyobe	9297 xaely	9347 xarel	9397 xedys	9447 xesuh	9497 xifer
9048 wojow	9098 wubaw	9148 wuosk	9198 wydak	9248 wyock	9298 xaerk	9348 xaruw	9398 xeeby	9448 xetar	9498 xifid
9049 wokem	9099 wuboj	9149 wuozu	9199 wydem	9249 wyolo	9299 xaevo	9349 xasax	9399 xeeco	9449 xetel	9499 xigay

9500 xigep	**9550** xityl	**9600** xoona	**9650** xushe	**9700** yalus	**9750** ybcre	**9800** ybofo	**9850** ycdaw	**9900** yctun	**9950** ydimi
9501 xigil	9551 xiuca	9601 xopat	9651 xusiv	9701 yamis	9751 ybdam	9801 yboke	9851 ycdex	9901 ycugi	9951 ydiru
9502 xigus	9552 xiudo	9602 xopoh	9652 xusny	9702 yamul	9752 ybdel	9802 ybold	9852 ycdis	9902 ycupo	9952 ydiso
9503 xihaz	9553 xiunt	9603 xopul	9653 xusok	9703 yanex	9753 ybdiv	9803 ybont	9853 ycdur	9903 ycurb	9953 ydjas
9504 xihex	9554 xiure	9604 xorer	9654 xusut	9704 yanil	9754 ybdro	9804 ybopu	9854 ycejo	9904 ycusa	9954 ydjet
9505 xihik	9555 xivam	9605 xorha	9655 xutay	9705 yanub	9755 ybdud	9805 yborb	9855 ycelf	9905 ycute	9955 ydjor
9506 xihum	9556 xivel	9606 xorim	9656 xutle	9706 yaofa	9756 ybebe	9806 ybovi	9856 ycene	9906 ycveh	9956 ydkav
9507 xihyp	9557 xivot	9607 xosey	9657 xutro	9707 yaoge	9757 ybend	9807 yboxa	9857 yceva	9907 ycvip	9957 ydkew
9508 xijab	9558 xiwan	9608 xoshi	9658 xuvop	9708 yaolm	9758 yberf	9808 ybpac	9858 ycexu	9908 ycvow	9958 ydkiz
9509 xijon	9559 xizeh	9609 xosma	9659 xuwas	9709 yaopo	9759 ybewo	9809 ybpeg	9859 ycfar	9909 ycvus	9959 ydlax
9510 xijuk	**9560** xizof	**9610** xosot	**9660** xuwep	**9710** yaorn	**9760** ybfan	**9810** ybpin	**9860** ycfez	**9910** ycwav	**9960** ydler
9511 xikac	9561 xoako	9611 xothe	9661 xuwib	9711 yaosk	9761 ybfew	9811 ybplu	9861 ycfik	9911 ycwec	9961 ydlot
9512 xikeg	9562 xoats	9612 xotib	9662 xuzun	9712 yaozu	9762 ybfod	9812 ybpom	9862 ycfog	9912 ycwon	9962 ydlup
9513 xikit	9563 xoaxa	9613 xotok	9663 yaaha	9713 yapef	9763 ybfri	9813 ybrad	9863 ycfux	9913 ydaci	9963 ydmaz
9514 xikul	9564 xoaze	9614 xotud	9664 yaalp	9714 yapru	9764 ybfum	9814 ybrec	9864 ycgaf	9914 ydado	9964 ydmep
9515 xikyn	9565 xobay	9615 xotyg	9665 yaamb	9715 yaraj	9765 ybgap	9815 ybrol	9865 ycged	9915 ydafu	9965 ydmin
9516 xilad	9566 xobom	9616 xowap	9666 yaark	9716 yarix	9766 ybges	9816 ybrub	9866 ycguz	9916 ydaga	9966 ydmok
9517 xilec	9567 xobri	9617 xowif	9667 yabaw	9717 yaser	9767 ybgiz	9817 ybsaf	9867 ychax	9917 ydalk	9967 ydmus
9518 xiliv	9568 xocho	9618 xoyab	9668 yaboj	9718 yaspa	9768 ybgof	9818 ybsed	9868 ychub	9918 ydash	9968 ydnab
9519 xilol	9569 xocip	9619 xoyes	9669 yacat	9719 yasto	9769 ybguc	9819 ybsig	9869 ycick	9919 ydbam	9969 ydnev
9520 xilub	**9570** xocly	**9620** xoyun	**9670** yacen	**9720** yasuh	**9770** ybhak	**9820** ybsoc	**9870** ycifi	**9920** ydbel	**9970** ydnig
9521 ximaf	9571 xodah	9621 xozar	9671 yacuf	9721 yatar	9771 ybhip	9821 ybsuz	9871 ycips	9921 ydbiv	9971 ydnuk
9522 ximed	9572 xodit	9622 xozeg	9672 yadew	9722 yatel	9772 ybhoz	9822 ybtav	9872 ycire	9922 ydbud	9972 ydola
9523 ximoc	9573 xodye	9623 xozis	9673 yadir	9723 yatif	9773 ybhux	9823 ybtem	9873 ycita	9923 ydcac	9973 ydomo
9524 ximuz	9574 xoeca	9624 xualt	9674 yadot	9724 yatom	9774 ybich	9824 ybtis	9874 ycker	9924 ydceg	9974 ydong
9525 xinag	9575 xoegy	9625 xuaum	9675 yadum	9725 yatug	9775 ybido	9825 ybtuf	9875 yckig	9925 ydclu	9975 ydose
9526 xineb	9576 xoels	9626 xubra	9676 yaeby	9726 yaula	9776 ybisi	9826 ybubs	9876 yckup	9926 ydcom	9976 ydour
9527 xinun	9577 xoeum	9627 xucla	9677 yaeco	9727 yaumo	9777 ybiza	9827 ybuck	9877 yclem	9927 yddan	9977 ydpad
9528 xiobi	9578 xoezo	9628 xucyt	9678 yaers	9728 yaupi	9778 ybjar	9828 ybuly	9878 yclin	9928 yddif	9978 ydpec
9529 xiodu	9579 xofak	9629 xudow	9679 yafaf	9729 yauvu	9779 ybjez	9829 ybuno	9879 yclob	9929 yddod	9979 ydpol
9530 xioft	**9580** xofle	**9630** xudre	**9680** yafli	**9730** yavep	**9780** ybjog	**9830** ybura	**9880** ycluf	**9930** yddut	**9980** ydpub
9531 xiogo	9581 xogeb	9631 xuelo	9681 yafos	9731 yavod	9781 ybkas	9831 ybuzi	9881 ycmal	9931 ydecu	9981 ydraf
9532 xioha	9582 xoglo	9632 xuerb	9682 yafre	9732 yaweg	9782 ybkek	9832 ybvag	9882 ycmen	9932 ydego	9982 ydred
9533 xiove	9583 xohas	9633 xuetz	9683 yaham	9733 yawov	9783 ybkid	9833 ybvil	9883 ycmos	9933 ydeja	9983 ydril
9534 xipaw	9584 xohed	9634 xufuc	9684 yaheb	9734 yaxan	9784 ybkor	9834 ybvox	9884 ycnam	9934 ydeli	9984 ydroc
9535 xipir	9585 xohuw	9635 xugah	9685 yahog	9735 yaxor	9785 ybkug	9835 ybvun	9885 ycnel	9935 ydeve	9985 ydruz
9536 xipoz	9586 xoila	9636 xuhef	9686 yahud	9736 yazak	9786 yblaw	9836 ybwir	9886 ycnit	9936 ydews	9986 ydsag
9537 xiput	9587 xoipe	9637 xuhim	9687 yaibo	9737 yazem	9787 ybleh	9837 ybwob	9887 ycnud	9937 ydfap	9987 ydseb
9538 xirob	9588 xoito	9638 xuhos	9688 yaile	9738 yazob	9788 yblit	9838 ycals	9888 ycocu	9938 ydfit	9988 ydsik
9539 xiruf	9589 xojam	9639 xukom	9689 yaink	9739 ybaik	9789 yblur	9839 ycamu	9889 ycoko	9939 ydfro	9989 ydsox
9540 xisal	**9590** xojop	**9640** xumym	**9690** yaixu	**9740** ybans	**9790** ybmax	**9840** ycank	**9890** ycoma	**9940** ydful	**9990** ydsun
9541 xiscu	9591 xokoy	9641 xunat	9691 yajip	9741 ybarg	9791 ybmer	9841 ycapi	9891 ycope	9941 ydgar	9991 ydtah
9542 xisen	9592 xokra	9642 xuofe	9692 yajok	9742 ybavu	9792 ybmib	9842 ycart	9892 ycres	9942 ydgez	9992 ydtim
9543 xiski	9593 xokuv	9643 xuohn	9693 yajun	9743 ybbal	9793 ybmov	9843 ycavo	9893 ycriv	9943 ydgog	9993 ydtob
9544 xisos	9594 xomet	9644 xuoty	9694 yakah	9744 ybben	9794 ybmut	9844 ycawa	9894 ycrof	9944 ydhat	9994 yduco
9545 xitap	9595 xomow	9645 xupri	9695 yaket	9745 ybbix	9795 ybnaz	9845 ycaxe	9895 ycseg	9945 ydhib	9995 yduda
9546 xitew	9596 xonaw	9646 xuret	9696 yakow	9746 ybbos	9796 ybnep	9846 ycbil	9896 ycsom	9946 ydhov	9996 ydume
9547 xitiz	9597 xonen	9647 xurin	9697 yakri	9747 ybcab	9797 ybnim	9847 ycboz	9897 yctak	9947 ydhuf	9997 yduph
9548 xitod	9598 xonuc	9648 xuroy	9698 yalaz	9748 ybcli	9798 ybnok	9848 yccod	9898 yctet	9948 ydiba	9998 yduvi
9549 xitur	9599 xooks	9649 xusca	9699 yalic	9749 ybcon	9799 ybnus	9849 yccum	9899 yctoc	9949 ydife	9999 ydvay

ydvef ydvid ydvoz ydwaw

SECTION C.

KEY ID:AL01154

SECTION C

0000 ydwem	**0050** yezon	**0100** yiloh	**0150** ylcir	**0200** ylriz	**0250** ymhow	**0300** ymugu	**0350** ynnak	**0400** yofis	**0450** ysbeg
0001 ydwis	0051 yezuf	0101 yilut	0151 ylcut	0201 ylron	0251 ymibu	0301 ymuko	0351 ynnec	0401 yogaw	0451 ysbif
0002 yeaka	0052 yiabe	0102 yimac	0152 yldac	0202 ylruk	0252 ymige	0302 ymule	0352 ynniz	0402 yoget	0452 ysblu
0003 yealo	0053 yiack	0103 yimeg	0153 yldeg	0203 ylsec	0253 ymind	0303 ymuna	0353 ynnof	0403 yogig	0453 ysbro
0004 yeamp	0054 yiaft	0104 yimif	0154 yldin	0204 ylsol	0254 ymirk	0304 ymvas	0354 ynnut	0404 yogro	0454 yscad
0005 yeanu	0055 yiagi	0105 yimom	0155 yldul	0205 ylsub	0255 ymivo	0305 ymvib	0355 ynocs	0405 yohan	0455 yscec
0006 yedru	0056 yiahu	0106 yinew	0156 ylefi	0206 ylted	0256 ymixa	0306 ymvor	0356 ynoka	0406 yohoc	0456 yscol
0007 yeefu	0057 yiaja	0107 yinik	0157 ylega	0207 yltip	0257 ymkam	0307 ynaam	0357 ynole	0407 yoims	0457 yscri
0008 yeerk	0058 yiarn	0108 yinol	0158 yleke	0208 yltox	0258 ymkel	0308 ynabs	0358 ynoro	0408 yoitz	0458 yscub
0009 yefal	0059 yiasp	0109 yiobo	0159 ylelm	0209 yltuc	0259 ymkif	0309 ynajo	0359 ynoxi	0409 yojav	0459 ysdaf
0010 yefet	**0060** yiawo	**0110** yiofy	**0160** ylepu	**0210** ylubo	**0260** ymkro	**0310** ynake	**0360** ynric	**0410** yojib	**0460** ysded
0011 yefif	0061 yiban	0111 yiome	0161 yleth	0211 yluka	0261 ymkud	0311 ynalu	0361 ynsib	0411 yojod	0461 ysdil
0012 yefoc	0062 yibet	0112 yionc	0162 ylfam	0212 ylune	0262 ymlan	0312 ynarc	0362 ynsko	0412 yojum	0462 ysdoc
0013 yegeh	0063 yibiz	0113 yiopa	0163 ylfel	0213 ylvat	0263 ymlet	0313 ynavi	0363 ynstu	0413 yokex	0463 ysduz
0014 yegow	0064 yibod	0114 yiori	0164 ylfig	0214 ylvix	0264 ymlis	0314 ynbex	0364 ynswe	0414 yoklo	0464 yseba
0015 yegra	0065 yicap	0115 yiots	0165 ylfot	0215 ylvoc	0265 ymlod	0315 yncah	0365 yntan	0415 yokut	0465 ysech
0016 yehus	0066 yiciv	0116 yiped	0166 ylfud	0216 ylwal	0266 ymlum	0316 yncig	0366 yntop	0416 yolab	0466 ysedo
0017 yeins	0067 yicof	0117 yipil	0167 ylgan	0217 ymala	0267 ymmap	0317 yndet	0367 yntux	0417 yolil	0467 ysege
0018 yeipa	0068 yicuc	0118 yipoc	0168 ylgew	0218 ymano	0268 ymmes	0318 yndom	0368 ynual	0418 yolop	0468 yseki
0019 yejaz	0069 yidar	0119 yirag	0169 ylgic	0219 ymare	0269 ymmiz	0319 ynebi	0369 ynubu	0419 yoluk	0469 ysemp
0020 yejec	**0070** yidez	**0120** yireb	**0170** ylgod	**0220** ymasi	**0270** ymmof	**0320** yneft	**0370** ynude	**0420** yomew	**0470** ysent
0021 yejol	0071 yidip	0121 yirid	0171 ylgum	0221 ymatu	0271 ymmuc	0321 ynelk	0371 ynufi	0421 yonin	0471 ysfag
0022 yejud	0072 yidog	0122 yirox	0172 ylhet	0222 ymawl	0272 ymnac	0322 yneme	0372 ynuor	0422 yonos	0472 ysfeb
0023 yekik	0073 yidux	0123 yirun	0173 ylhor	0223 ymbad	0273 ymneg	0323 ynena	0373 ynvis	0423 yopla	0473 ysfir
0024 yekof	0074 yiefo	0124 yisaw	0174 ylhun	0224 ymbec	0274 ymniv	0324 ynesh	0374 ynvon	0424 yopof	0474 ysfox
0025 yelak	0075 yielv	0125 yiten	0175 yliku	0225 ymbik	0275 ymnom	0325 ynfra	0375 ynwaf	0425 yopre	0475 ysfun
0026 yelev	0076 yiemu	0126 yitha	0176 ylima	0226 ymbol	0276 ymoce	0326 ynfuf	0376 ynwel	0426 yorah	0476 ysgat
0027 yenad	0077 yierg	0127 yiuba	0177 ylise	0227 ymbub	0277 ymodo	0327 yngaz	0377 ynwog	0427 yorog	0477 ysgef
0028 yenir	0078 yiese	0128 yiufe	0178 ylkar	0228 ymcaf	0278 ymoli	0328 yngli	0378 yoand	0428 yosak	0478 ysgid
0029 yenul	0079 yifas	0129 yiuku	0179 ylkib	0229 ymced	0279 ymoms	0329 yngoc	0379 yobat	0429 yoshu	0479 yshaw
0030 yeoso	**0080** yifix	**0130** yiuni	**0180** ylkog	**0230** ymcoc	**0280** ymonz	**0330** yngre	**0380** yobeb	**0430** yosme	**0480** yshut
0031 yeozi	0081 yifor	0131 yiuzo	0181 ylmaj	0231 ymcuz	0281 ymoth	0331 yngud	0381 yobir	0431 yotag	0481 ysibi
0032 yepha	0082 yifug	0132 yivak	0182 ylmeh	0232 ymdag	0282 ymowa	0332 ynhez	0382 yobow	0432 yotek	0482 ysida
0033 yeple	0083 yigav	0133 yivem	0183 ylmik	0233 ymdeb	0283 ymoxu	0333 ynhid	0383 yobul	0433 yotho	0483 ysing
0034 yeroj	0084 yigim	0134 yivic	0184 ylmop	0234 ymdid	0284 ympax	0334 ynhob	0384 yocal	0434 yotuz	0484 ysith
0035 yeruw	0085 yigop	0135 yivul	0185 ylnax	0235 ymdox	0285 ymper	0335 ynimy	0385 yocef	0435 yousi	0485 ysjak
0036 yeske	0086 yigur	0136 yiweh	0186 ylner	0236 ymdun	0286 ympic	0336 ynini	0386 yocim	0436 yowam	0486 ysjem
0037 yesli	0087 yihab	0137 yladu	0187 ylnif	0237 ymeld	0287 ympov	0337 ynisk	0387 yocot	0437 yower	0487 ysjob
0038 yesov	0088 yihek	0138 ylafa	0188 ylnov	0238 ymepi	0288 ympup	0338 ynizo	0388 yodad	0438 yowiz	0488 ysjuf
0039 yespu	0089 yihig	0139 ylale	0189 ylnup	0239 ymeru	0289 ymrav	0339 ynkaw	0389 yodec	0439 yoxem	0489 yskal
0040 yetaf	**0090** yihon	**0140** ylani	**0190** ylode	**0240** ymete	**0290** ymrex	**0340** ynkem	**0390** yodol	**0440** ysabo	**0490** ysken
0041 yetes	0091 yihup	0141 ylaph	0191 ylomi	0241 ymfat	0291 ymrip	0341 ynkot	0391 yodub	0441 ysacs	0491 yskos
0042 yeudu	0092 yikaz	0142 ylarx	0192 ylops	0242 ymfeh	0292 ymrot	0342 ynkuk	0392 yoeda	0442 ysald	0492 yslam
0043 yeums	0093 yikep	0143 ylato	0193 ylorf	0243 ymfim	0293 ymrul	0343 ynlav	0393 yoemb	0443 ysarm	0493 yslel
0044 yeunk	0094 yikin	0144 ylbak	0194 ylotu	0244 ymfoz	0294 ymsar	0344 ynlep	0394 yoeni	0444 ysate	0494 yslix
0045 yeuri	0095 yikok	0145 ylbem	0195 ylpaz	0245 ymgak	0295 ymsez	0345 ynlir	0395 yoeps	0445 ysauk	0495 yslon
0046 yeuto	0096 yikus	0146 ylbid	0196 ylpep	0246 ymgem	0296 ymsin	0346 ynlok	0396 yoerz	0446 ysawn	0496 yslud
0047 yeuve	0097 yilal	0147 ylbob	0197 ylpit	0247 ymgob	0297 ymsog	0347 ynmat	0397 yoetu	0447 ysaxi	0497 ysman
0048 yevig	0098 yilex	0148 ylbuf	0198 ylrab	0248 ymguf	0298 ymtus	0348 ynmod	0398 yoexo	0448 ysaza	0498 ysmex
0049 yezac	0099 yilib	0149 ylcaw	0199 ylrev	0249 ymhil	0299 ymudi	0349 ynmup	0399 yofed	0449 ysbac	0499 ysmoh

0500 ysmum	**0550** ytero	**0600** ytren	**0650** yugle	**0700** yuwax	**0750** yxits	**0800** yxwid	**0850** zakeg	**0900** zavij	**0950** zekaf
0501 ysnap	0551 ytest	0601 ytros	0651 yugun	0701 yuwig	0751 yxkag	0801 zaami	0851 zakit	0901 zavot	0951 zeked
0502 ysnes	0552 ytewe	0602 ytsam	0652 yuhah	0702 yuwom	0752 yxkeb	0802 zaanz	0852 zakul	0902 zawan	0952 zekli
0503 ysobu	0553 ytfaz	0603 ytsel	0653 yuheg	0703 yuzaf	0753 yxkil	0803 zaapa	0853 zakyn	0903 zayat	0953 zekoc
0504 ysols	0554 ytfep	0604 ytsho	0654 yuhol	0704 yuzed	0754 yxkun	0804 zaask	0854 zalad	0904 zazeh	0954 zelag
0505 ysond	0555 ytfin	0605 ytsif	0655 yuhur	0705 yuzib	0755 yxlef	0805 zaath	0855 zalec	0905 zazof	0955 zeleb
0506 ysorc	0556 ytfly	0606 ytsud	0656 yuico	0706 yuzoc	0756 yxlig	0806 zaaxo	0856 zaliv	0906 zeabi	0956 zeliz
0507 ysoto	0557 ytfok	0607 yttat	0657 yuifa	0707 yxaba	0757 yxloz	0807 zabar	0857 zalol	0907 zeact	0957 zelox
0508 ysoze	0558 ytfus	0608 yttez	0658 yuils	0708 yxach	0758 yxluv	0808 zabez	0858 zalub	0908 zeady	0958 zelun
0509 yspar	0559 ytgab	0609 yttig	0659 yuitu	0709 yxaim	0759 yxmem	0809 zabim	0859 zamaf	0909 zeage	0959 zelyt
0510 yspez	**0560** ytgev	**0610** ytufo	**0660** yuive	**0710** yxaju	**0760** yxmip	**0810** zably	**0860** zamed	**0910** zeama	**0960** zemah
0511 yspik	0561 ytgis	0611 ytuma	0661 yuizi	0711 yxant	0761 yxmob	0811 zabog	0861 zamoc	0911 zeapo	0961 zemef
0512 yspog	0562 ytgon	0612 ytund	0662 yujef	0712 yxase	0762 yxmuf	0812 zabux	0862 zamuz	0912 zebav	0962 zemic
0513 ysrib	0563 ythap	0613 ytuse	0663 yujid	0713 yxazo	0763 yxnal	0813 zacas	0863 zanag	0913 zebij	0963 zenay
0514 yssav	0564 ythof	0614 ytuti	0664 yujot	0714 yxbax	0764 yxnek	0814 zacin	0864 zaneb	0914 zeble	0964 zenem
0515 yssim	0565 ythuc	0615 ytvan	0665 yujup	0715 yxber	0765 yxnix	0815 zacor	0865 zanis	0915 zeboh	0965 zenid
0516 yssop	0566 ytibs	0616 ytvik	0666 yukak	0716 yxblo	0766 yxobs	0816 zacug	0866 zanox	0916 zebur	0966 zenob
0517 ystab	0567 ytice	0617 ytvum	0667 yukeh	0717 yxbup	0767 yxoda	0817 zacyb	0867 zanun	0917 zecax	0967 zenuf
0518 ystit	0568 ytiod	0618 ytwar	0668 yukob	0718 yxcaz	0768 yxofi	0818 zadav	0868 zaobi	0918 zecer	0968 zeoak
0519 ystow	0569 ytipi	0619 ytwes	0669 yukuf	0719 yxcep	0769 yxojo	0819 zadej	0869 zaodu	0919 zecov	0969 zeofu
0520 ysube	**0570** ytirl	**0620** ytwot	**0670** yulat	**0720** yxcro	**0770** yxoku	**0820** zadib	**0870** zaoft	**0920** zecup	**0970** zeoly
0521 ysuja	0571 ytjac	0621 yuade	0671 yulov	0721 yxcus	0771 yxowl	0821 zadop	0871 zaogo	0921 zedaz	0971 zeomp
0522 ysult	0572 ytjeg	0622 yuaku	0672 yulul	0722 yxdab	0772 yxpam	0822 zaebs	0872 zaoha	0922 zedep	0972 zeone
0523 ysumy	0573 ytjuw	0623 yuamo	0673 yumab	0723 yxdev	0773 yxpel	0823 zaegi	0873 zaove	0923 zedix	0973 zeort
0524 ysunz	0574 ytkad	0624 yuapt	0674 yumev	0724 yxdic	0774 yxpiv	0824 zaema	0874 zaoys	0924 zedok	0974 zeosi
0525 ysver	0575 ytkec	0625 yuars	0675 yumim	0725 yxdon	0775 yxpot	0825 zaeno	0875 zapaw	0925 zeedi	0975 zeowo
0526 ysvov	0576 ytkir	0626 yuava	0676 yumon	0726 yxduk	0776 yxpud	0826 zaepe	0876 zapir	0926 zeemo	0976 zepig
0527 ysvup	0577 ytkol	0627 yubaz	0677 yumuk	0727 yxebu	0777 yxran	0827 zaert	0877 zapoz	0927 zeeng	0977 zerat
0528 yswaz	0578 ytkub	0628 yubin	0678 yunan	0728 yxefa	0778 yxrew	0828 zaezu	0878 zaput	0928 zeerl	0978 zerez
0529 yswet	0579 ytlaf	0629 yubok	0679 yunod	0729 yxele	0779 yxrik	0829 zafax	0879 zarak	0929 zeesy	0979 zerow
0530 yswod	**0580** ytloc	**0630** yubre	**0680** yunum	**0730** yxens	**0780** yxrod	**0830** zafer	**0880** zarem	**0930** zeewa	**0980** zeryn
0531 ytabu	0581 ytluz	0631 yubus	0681 yuogi	0731 yxeph	0781 yxrum	0831 zafid	0881 zarig	0931 zeexe	0981 zesew
0532 ytaca	0582 ytmag	0632 yucam	0682 yuono	0732 yxerd	0782 yxsap	0832 zagay	0882 zarob	0932 zefip	0982 zesil
0533 ytads	0583 ytmeb	0633 yucel	0683 yuork	0733 yxfac	0783 yxses	0833 zagep	0883 zaruf	0933 zeflu	0983 zesod
0534 ytago	0584 ytmil	0634 yucud	0684 yuosa	0734 yxfeg	0784 yxsiz	0834 zagil	0884 zasal	0934 zegac	0984 zesum
0535 ytalm	0585 ytmun	0635 yudal	0685 yuovu	0735 yxfib	0785 yxsof	0835 zahaz	0885 zascu	0935 zegeg	0985 zetas
0536 ytane	0586 ytnah	0636 yuden	0686 yupap	0736 yxfom	0786 yxsuc	0836 zahex	0886 zasen	0936 zegin	0986 zetey
0537 ytawk	0587 ytnef	0637 yudik	0687 yupes	0737 yxfru	0787 yxtaw	0837 zahik	0887 zaski	0937 zegom	0987 zetik
0538 ytbek	0588 ytnib	0638 yudos	0688 yupiz	0738 yxgad	0788 yxtex	0838 zahum	0888 zasos	0938 zegyl	0988 zetol
0539 ytbli	0589 ytnoz	0639 yuecy	0689 yupro	0739 yxgec	0789 yxthi	0839 zahyp	0889 zatap	0939 zehad	0989 zetut
0540 ytbor	**0590** ytoba	**0640** yueka	**0690** yupuc	**0740** yxgif	**0790** yxtor	**0840** zaide	**0890** zatew	**0940** zehot	**0990** zeufa
0541 ytbug	0591 ytord	0641 yuelb	0691 yurit	0741 yxgol	0791 yxuga	0841 zaili	0891 zatiz	0941 zehug	0991 zeunc
0542 ytcur	0592 ytote	0642 yuenu	0692 yusas	0742 yxgub	0792 yxumi	0842 zaimu	0892 zatod	0942 zeicy	0992 zeurg
0543 ytdax	0593 ytown	0643 yuesi	0693 yusek	0743 yxhit	0793 yxurn	0843 zaipo	0893 zatur	0943 zeigi	0993 zeust
0544 ytder	0594 ytoxo	0644 yufad	0694 yuspi	0744 yxhul	0794 yxutu	0844 zaist	0894 zatyl	0944 zeinz	0994 zeuze
0545 ytdiz	0595 ytpak	0645 yufec	0695 yusug	0745 yxibe	0795 yxvar	0845 zajab	0895 zauca	0945 zeism	0995 zevap
0546 ytdov	0596 ytpem	0646 yuflo	0696 yutil	0746 yxica	0796 yxvez	0846 zajif	0896 zaudo	0946 zejeh	0996 zeves
0547 ytdup	0597 ytplo	0647 yufub	0697 yuvav	0747 yxifo	0797 yxvin	0847 zajon	0897 zaunt	0947 zejim	0997 zeviv
0548 ytemi	0598 ytpuf	0648 yugag	0698 yuvet	0748 yxigu	0798 yxvog	0848 zajuk	0898 zaure	0948 zejoy	0998 zewit
0549 ytepa	0599 ytral	0649 yugir	0699 yuvif	0749 yximp	0799 yxwat	0849 zakac	0899 zavam	0949 zejub	0999 zewor

zexel zeyar zeyex zeyog

[redacted]

[redacted]

SECTION C

1000 zezek	**1050** zilaf	**1100** ziwar	**1150** zokav	**1200** zovay	**1250** zugim	**1300** zurox
1001 zezir	1051 ziled	1101 ziwes	1151 zokew	1201 zovef	1251 zugop	1301 zurun
1002 ziabu	1052 zilip	1102 ziwot	1152 zokiz	1202 zovid	1252 zugur	1302 zusaw
1003 ziaca	1053 ziloc	1103 zizog	1153 zokry	1203 zovoz	1253 zugyn	1303 zusef
1004 ziads	1054 ziluz	1104 zizul	1154 zolax	1204 zowaw	1254 zuhab	1304 zusit
1005 ziafi	1055 zilyn	1105 zoaci	1155 zoler	1205 zowem	1255 zuhig	1305 zusoz
1006 ziago	1056 zimag	1106 zoado	1156 zolot	1206 zowis	1256 zuhon	1306 zusuv
1007 zialm	1057 zimeb	1107 zoafu	1157 zolup	1207 zoyak	1257 zuhup	1307 zusys
1008 ziane	1058 zimil	1108 zoaga	1158 zomaz	1208 zozal	1258 zuimo	1308 zuten
1009 ziaty	1059 zimox	1109 zoalk	1159 zomep	1209 zozen	1259 zuiny	1309 zutha
1010 ziawk	**1060** zimun	**1110** zoash	**1160** zomin	**1210** zozip	**1260** zuipu	**1310** zutry
1011 zibas	1061 zimyc	1111 zobam	1161 zomok	1211 zozos	1261 zuiti	1311 zutuk
1012 zibek	1062 zinah	1112 zobel	1162 zomus	1212 zozug	1262 zuiwa	1312 zuvak
1013 zibli	1063 zinef	1113 zobiv	1163 zonab	1213 zuabe	1263 zujah	1313 zuvem
1014 zibor	1064 zinib	1114 zoboy	1164 zonev	1214 zuack	1264 zujer	1314 zuvic
1015 zibug	1065 zinoz	1115 zobud	1165 zonig	1215 zuaft	1265 zujis	1315 zuvul
1016 zicav	1066 zinyl	1116 zocac	1166 zonon	1216 zuagi	1266 zukaz	1316 zuweh
1017 zicop	1067 zioci	1117 zoceg	1167 zonuk	1217 zuahu	1267 zukep	1317 zuyle
1018 zicur	1068 zioky	1118 zoclu	1168 zonyp	1218 zuaja	1268 zukin	1318 zuyot
1019 zicyg	1069 ziord	1119 zocom	1169 zoody	1219 zuamy	1269 zukok	1319 zuyub
1020 zidax	**1070** ziote	**1120** zocys	**1170** zoola	**1220** zuarn	**1270** zukus	**1320** zuzam
1021 zider	1071 ziown	1121 zodan	1171 zoomo	1221 zuasp	1271 zukym	1321 zuzel
1022 zidov	1072 zioxo	1122 zodey	1172 zoong	1222 zuawo	1272 zulal	1322 zuzoy
1023 zidup	1073 zipak	1123 zodif	1173 zoose	1223 zuban	1273 zulex	
1024 ziemi	1074 zipem	1124 zodod	1174 zopad	1224 zubet	1274 zulib	
1025 ziepa	1075 zipid	1125 zodut	1175 zopec	1225 zubiz	1275 zuloh	
1026 ziero	1076 ziplo	1126 zoecu	1176 zopij	1226 zubod	1276 zulut	
1027 ziewe	1077 zipry	1127 zoego	1177 zopol	1227 zubum	1277 zulyr	
1028 zifaz	1078 zipuf	1128 zoeja	1178 zopub	1228 zubyc	1278 zumac	
1029 zifep	1079 ziral	1129 zoeli	1179 zopyr	1229 zucap	1279 zumeg	
1030 zifin	**1080** ziren	**1130** zoerx	**1180** zoraf	**1230** zuces	**1280** zumif	
1031 zifly	1081 ziros	1131 zoeve	1181 zored	1231 zucof	1281 zumom	
1032 zifok	1082 zirut	1132 zofap	1182 zoroc	1232 zucuc	1282 zunew	
1033 zifus	1083 zisam	1133 zofes	1183 zoruz	1233 zucyx	1283 zunik	
1034 zigab	1084 zisho	1134 zofit	1184 zosag	1234 zudar	1284 zunol	
1035 zigev	1085 zisif	1135 zofro	1185 zoseb	1235 zudez	1285 zuobo	
1036 zigon	1086 zisud	1136 zoful	1186 zosik	1236 zudip	1286 zuofy	
1037 zihap	1087 zisyk	1137 zogar	1187 zosox	1237 zudog	1287 zuome	
1038 zihew	1088 zitez	1138 zogez	1188 zosun	1238 zudux	1288 zuonc	
1039 zihof	1089 zitig	1139 zogog	1189 zosyd	1239 zuedy	1289 zuopa	
1040 zihuc	**1090** zitoy	**1140** zohat	**1190** zotah	**1240** zuefo	**1290** zuori	
1041 zihyd	1091 zitru	1141 zohib	1191 zotim	1241 zuelv	1291 zuots	
1042 zijac	1092 ziufo	1142 zohov	1192 zotob	1242 zuemu	1292 zupaf	
1043 zijeg	1093 ziuma	1143 zohuf	1193 zouco	1243 zuerg	1293 zuped	
1044 zijuw	1094 ziund	1144 zohyg	1194 zouda	1244 zuese	1294 zupil	
1045 zikad	1095 ziuse	1145 zoiba	1195 zouky	1245 zufas	1295 zupoc	
1046 zikec	1096 ziuti	1146 zoife	1196 zoume	1246 zufix	1296 zupuz	
1047 zikir	1097 zivan	1147 zoimi	1197 zouph	1247 zufor	1297 zurag	
1048 zikol	1098 zivik	1148 zoiru	1198 zouvi	1248 zufug	1298 zureb	
1049 zikub	1099 zivum	1149 zojet	1199 zouzu	1249 zugav	1299 zurid	

TERMINATIONAL
ORDER.

[redacted]

Terminational Order.

A

ac ba do eg fe ib ky om pu ri uw yx	aba
ab bi co da ev gu ij le my uk ys	eba
al by en ge if os ru uj va yd zo	iba
af ed ga hy ig ko uz vu yt	oba
ca it ly ne ob so uf yi	uba
ju la ok ro us we	yba
be ed ga hy ig ko oc uz vu yt zi	aca
he in oq xo	eca
ac ba do eg fe ib ky om pu uw yx	ica
bo cy de ef mu oz ra si uv	oca
aq oh xu	sca
ag ce dy eb ik ox un xi za	uca
ax je no ov ta up	yca
am el fo iv ke nu oj sa ud	ada
ex hu ir ot pe ry vi wa yo	eda
co da le on uk ys	ida
ac ba do eg fe ib ky om pu ri uw yx	oda
by ge if ji uj va yd zo	uda
eh fi go ip ka lu od py se um yl	afa
ba eg ib ky om ri uw yx	efa
ax er iw je no ov sy up yu	ifa
aw et ho ic li uq vy wu xe ya	ofa
fa hi iq ny ol ub vo ze	ufa
en ge if ji ru uj yd zo	aga
an eh fi go ka lu od py se um yl	ega
av du eq fy iz ja op re to ur	iga
as ek fu im jo or pa te ug wi	oga
ac ba do eg fe ib ky om ri uw yx.	uga
aw et ho ic li uq vy wu xe ya	aha
av du eq fy ja op re to ur	eha
ex ot ry ul wa	iha
ag bu ce dy eb ik ox un xi za	oha
ay di ej ix ow ut xa ye	pha
he in oq xo	rha
ak em it ly ne ob so uf yi zu	tha
aq ey oh	uha
ak ca em it ly ne ob uf yi zu	aja
by en ge if os ru va yd so	eja
eh fi go ip ka lu od py se um	oja
ab bi co da ev ij le my on uk ys	uja
ay di ej ix ow ut xa ye	aka
ax er iw je mi no ov sy ta up yu	eka
aj bo cy de ef il mu ra uv	ika
at ew ha is ki oy tu wo yn	oka
an eh fi go ip ka od py se um yl	uka
ap gy lo me na of pi uc ym	ala
aq xu	cla
ar cu ez ma og po ti ux ve wy	ela
he in oq xo	ila
al by en if ji os ru uj va yd zo	ola
ah ex hu ir pe ry ul vi wa yo	pla
at ha is ki oy tu uh	sla
et ho ic li uq vy xe ya	ula
co da ev gu le on uk	yla
ec fa hi iq ny ol su ub vo ze	ama
bu ce dy eb ik mo ox un xi za	ema
eh go ip ka lu od py um yl	ima
az ep ju la ni ok ro ty us we yc	oma

A

he in oq xo	sma
af be ed ga hy ig ko oc uz yt zi	uma
cu ez id ma og ti ux ve wy	ana
ew ha is ki oy tu uh wo yn	ena
bu ce dy eb ik mo ox un	ina
he in oq xo	ona
ap ih lo me na of pi uc ym	una
bu ce dy eb ik mo ox un xi za	apa
af be ed ga ig ko oc vu yt zi	epa
ay ej ix ow ut xa ye	ipa
ak ca em it ne ob so uf yi zu	opa
aw ho ic li uq vy wu xe ya	spa
ar ez id ma og po ti ux ve wy	upa
as ek fu jo or pa te ug	ara
aq ey oh xu	bra
av ci du eq fy iz ja op re to ur	cra
at ew ha is ki oy tu uh wo yn	fra
ay di ej ix ow ut xa ye	gra
he in oq xo	kra
ah ir ot pe ry ul vi wa	ora
el fo gi iv oj ud yb	ura
ap es ku lo me na of uc	yra
ez id ma og po ux wy	isa
ax er iw je mi ov sy ta yu	osa
az ep la ni ro ty we yc	usa
av ci du eq fy ja op to ur	ata
am el fo gi iv ke nu ud	eta
az ep ju la ok ro ty us we yc	ita
ar cu ez id ma og po ti ux ve wy	ota
ax er iw je mi no ov sy ta up	uta
iw je mi no sy ta yu	ava
az ep ju ok ro ty we yc	eva
ek fu jo or pa te ug	iva
ay di ej ix ow qu ut xa	ova
az ep ju la ni ok ro ty us we yc	awa
ad ec fa hi ny ol su ub vo ze	ewa
ak ca em it ly ne ob so uf zu	iwa
ap es gy ih ku lo me na of pi uc ym	owa
he in oq xo	axa
as ek fu im jo or pa te ug wi	exa
ap gy ih ku lo na of uc ym	ixa
am el fo gi iv ke nu oj sa ud yb	oxa
ay di ej ix ow ut xa	uxa
aj bo cy de ef il mu oz ra si uv	aya
ab bi co da ev ij le my on uk ys	aza
am fo iv ke nu oj sa ud yb	iza
as ek im jo pa te ug wi	uza

B

av ci eq fy iz ja op re to ur	bab
am el fo iv nu oj sa ud yb	cab
ac do eg ib om pu uw yx	dab
ad ec fa hi iq ny ol su ub vo	fab
af ed hy ig ko oc uz vu yt zi	gab
ak ca em it ly ne ob so uf yi zu	hab
ag bu ce eb ik mo ox un xi za	jab
aj bo cy de ef mu oz ra uv	kab
ah ex ir ot pe ry ul vi wa yo	lab

B

er je mi ov sy yu	mab
al en ge if ji ru yd zo	nab
ar ez id og ux ve wy	pab
an eh go ip ka lu od py se um yl	rab
ek im jo or te ug	sab
ij my on uk ys	tab
ay di ej ix ow qu xa	wab
in xo	yab
ap ku lo na of pi uc	zab
ah ex hu ir ot ry ul vi wa yo	beb
as ek fu im pa te ug wi	ceb
ap es gy ih ku lo me na of pi uc ym	deb
ab bi co da ev gu ij le my on uk ys	feb
he in xo	geb
aw ho ic li uq vy wu ya	heb
ar cu id ma og po ti ux	jeb
ac ba do eg fe ib ky om pu ri uw yx	keb
ec iq ny ol su ub vo ze	leb
af be ed ga hy ig ko oc uz vu yt zi	meb
ag bu ce dy eb un xi za	neb
aj bo cy de ef il mu oz ra si uv	peb
ak ca em it ly ne ob so uf yi zu	reb
al en ge if ji os ru uj va yd zo	seb
av ci du eq fy iz ja op re	web
an eh go ip ka lu	yeb
cu ez id ma og po ti ux ve wy	bib
ad ec fa hi iq ny ol su ub vo	cib
ag bu ce dy eb un xi za	dib
ac ba do eg ib ky om pu ri uw yx	fib
aj bo cy de ef il mu oz ra si uv	gib
al by en ge if ji ru uj va yd zo	hib
ah ex ir ul vi wa yo	jib
an eh fi go ip ka lu od py se um yl	kib
ak ca em it ly ne ob so uf yi zu	lib
am fo gi iv ke nu oj sa ud yb	mib
af be ed ga hy ig ko oc uz vu yt zi	nib
av ci du eq fy iz ja op re to ur	pib
bi co da ev gu ij le on ys	rib
at ew ha is ki uh yn	sib
he in oq xo	tib
ap es gy ih ku lo me na of pi uc ym	vib
aq ey oh xu	wib
er iw je mi no ov ta up yu	zib
aj bo cy de ef il mu ra si uv	alb
ax er iw je mi no ov sy ta up yu	elb
aw et ho ic li uq vy wu xe ya	amb
ah ex hu ir ot pe ry ul vi wa yo	emb
an eh fi ip lu od py se um yl	bob
av du eq fy iz op re ur	cob
ar cu ez id ma og po ti ux ve wy	dob
ap es gy ih ku lo na of pi uc ym	gob
ew ha ki tu yn	hob
bi co da ev gu le on uk ys	job
ax er iw je mi no ov sy ta up yu	kob
az ep ju la ni ok ro ty we yc	lob
ac ba do eg ib ky om pu uw yx	mob
ec iq ny ol ub vo ze	nob
ag bu ce dy eb ik mo ox un xi za	rob

Terminational Order.

B

aj cy ef il mu oz ra si uv	sob
al by en if ji os ru va yd zo	tob
am el fo gi iv ke nu oj sa ud yb	wob
aw et ho li wu xe ya	zob
aq ey oh xu	erb
am el fo gi iv ke nu oj sa ud yb	orb
az ep la ni ok ro ty us we yc	urb
ap gy ih ku lo na of pi uc ym	bub
bi co da ev gu ij le on uk ys	cub
ah ex ir ot pe ry ul vi wa yo	dub
ax iw je mi no ov sy ta up yu	fub
ac ba do eg fe ib ky om pu ri uw yx	gub
az ep ju la ni ro ty we yc	hub
ec ol su ub vo ze	jub
af be ed ga hy ig ko oc uz vu yt zi	kub
ag bu ce dy eb ik mo ox un xi za	lub
aw et ho ic li uq vy wu xe ya	nub
al by en ge if ji os ru uj va yd zo	pub
am el fo iv ke nu oj sa ud yb	rub
an eh go ip ka od py se um yl	sub
as ek fu jo or pa te ug wi	tub
ak ca em it ne ob so uf zu	yub
av ci du ja op re to	zub
aj cy de ef il mu oz ra si uv	byb
ag bu ce dy eb ik mo ox xi za	cyb
ab bi co da ev gu le on uk	gyb
ay di ej ix ow qu ut xa	hyb
an eh fi go ip ka lu od py se um	lyb
az ep ju la ok ro us we	ryb
aw et ho ic li wu xe	tyb

C

bi co da ev gu ij le uk ys	bac
al by en ge if ji os ru uj va yd zo	cac
an eh fi go ip ka lu od py se um yl	dac
ac ba do eg ib ky om pu ri uw yx	fac
ad ec fa hi ny ol su vo ze	gac
at ha ki tu wo	hac
af ed ga ko oc vu yt zi	jac
ag bu ce dy eb ik mo ox un xi za	kac
aj bo cy de ef mu oz ra si uv	lac
ak ca em it ly ne ob uf yi zu	mac
ap es gy ih ku lo me na of pi uc ym	nac
el fo gi iv nu oj sa ud yb	pac
av ci du eq fy iz ja op re to ur	rac
az ep ju ni ok ro ty we	sac
ax er iw je mi no ov sy up	tac
as ek fu im jo or pa te ug wi	vac
ar cu ez id ma og po ti ux ve wy	wac
aw ho wu xe	yac
di ix ow xa ye	zac
aw et ic	aec
ap gy ih ku lo na of pi uc ym	bec
ab bi co da ev gu ij le my on uk ys	cec
ah ex hu ir ot pe ry ul vi wa yo	dec
ax iw je mi no ov sy ta up yu	fec
ac ba do eg fe ib ky om pu ri uw yx	gec
av ci du eq fy ja op re to ur	hec
di ix ow ut xa ye	jec
af be ed ga hy ig ko oc uz vu yt zi	kec
ag bu ce dy eb ik mo ox un xi za	lec
aj cy de ef il mu oz ra si uv	mec
at ew ha is oy tu uh wo yn	nec
al by en ge if ji os ru uj va yd zo	pec
am el fo iv ke nu oj sa ud yb	rec
an eh go ip ka od py se um yl	sec
as ek fu jo or pa te ug wi	tec
ar cu ez id ma og po ti ux ve wy	vec
az ep ju la ni ok ro ty us we yc	wec
aj bo cy de ef il mu oz ra si uv	cic
ac ba do eg fe ib ky om pu ri uw yx	dic
ez id ma og po ti ux ve wy	fic
an fi ip ka lu od py se um yl	gic
ep ju la ro ty	hic
aw et ho ic li uq vy wu xe ya	lic
ad fa hi iq ny su ub ze	mic
ab bi da ev gu ij le my on	nic
ap ih ku lo na of pi uc ym	pic
ew ha is tu uh wo yn	ric
ay di ej ix ow qu ut xa	tic
ak ca em it ly ne ob so uf yi zu	vic
as ek fu im jo or pa te ug wi	wic
am fo iv ke nu sa ud	zic
ar cu ez id ma og po ti ux ve wy	alc
ak ca em it ly ne ob so uf yi zu	onc
ad ec fa hi iq ny ol ub vo ze	unc
as ek fu im jo or pa te ug wi	boc
ap es gy lo me na of uc ym	coc
ab co da ev gu ij le my on uk ys	doc
ay di ej ix ow qu ut xa ye	foc
at ew ha is ki oy tu uh wo yn	goc
ex ir ot pe ry ul vi wa yo	hoc
ac ba do eg fe ib om pu ri uw	joc
ad ec fa hi iq ny ol su ub vo ze	koc
af be ed ga hy ig ko oc uz vu yt zi	loc
ag bu ce dy eb ik mo ox un xi za	moc
aj bo cy de ef il mu oz ra si uv	noc
ak em it ly ob so uf yi zu	poc
al en ge if ru uj yd zo	roc
am el fo gi iv ke nu oj sa ud yb	soc
az ep ju la ni ok ro ty us we yc	toc
an eh fi go ip ka lu od py se um yl	voc
ar cu id ma og po ve	yoc
ax er je mi no ov ta up yu	zoc
at ew ha is ki oy tu uh yn	arc
bi co da ev gu ij le on uk ys	orc
ag dy eb ik ox un	auc
aj bo cy de ef il mu oz ra si uv	buc
ak ca em it ly ne ob so uf yi zu	cuc
av ci du eq fy iz ja op re to ur	duc
at ew is oy	euc
aq ey oh xu	fuc
am fo iv ke nu oj sa ud yb	guc
af be ed ga hy ig ko oc uz vu yt zi	huc
ar cu ez id ma og po ti ux ve wy	kuc
as ek fu im jo or pa te ug wi	luc
ap es gy ih ku lo me na of pi uc ym	muc
he in oq xo	nuc
ax iw je mi no ov sy ta up yu	puc
az ep ju la ni ok ro ty us we	ruc
ac ba do eg ib ky om pu uw yx	suc
an eh fi go ip ka od py se um yl	tuc
ix ow ut xa	yuc
ah ex hu ir ot pe vi wa	zuc
ak ca em ly ne ob so uf zu	byc
av ci du fy iz ja op re to ur	cyc
aj bo cy de ef il mu oz ra si uv	dyc
ab bi co da ev gu ij le my on uk	hyc
at ew ha is ki oy tu uh wo	lyc
af be ed ga hy ig ko oc uz vu zi	myc
az ep ju la ni ok ro ty us we	nyc
ar cu ez id ma ti ux ve	pyc
as ek fu im jo or pa te ug wi	ryc
di ix ow ut xa	syc

D

ap gy ih ku me of pi uc ym	bad
bi co da ev gu ij le uk ys	cad
ah ex ir ot pe ry ul vi wa yo	dad
oh	ead
iw je mi no ov sy ta yu	fad
ac ib ky om uw yx	gad
ad fa hi iq ny ol su ub vo ze	had
at ew ha is ki tu uh wo	jad
af be ed ga hy ig ko oc uz vu yt zi	kad
ag bu ce dy eb un xi za	lad
aj cy de ef il oz si uv	mad
ay ej ix qu ut xa ye	nad
en ge if ji ru uj yd zo	pad
am el iv ke nu oj sa ud yb	rad
an eh fi ip ka od py se um	sad
as jo ug	tad
ar cu ez id ma og po ti ux ve wy	vad
et ic li uq wu xe	wad
av ci du ja op re to ur	zad
as ek fu jo pa te ug wi	bed
ap es gy ih ku lo me na of pi uc ym	ced
ab da ev gu ij le my on uk ys	ded
ah ex ir ot pe ry ul vi wa yo	fed
az ep ju la ni ok ro ty we yc	ged
he oq xo	hed
ow ut	ied
ac ba do eg fe ib om pu ri uw	jed
ad ec fa hi iq ny ol su ub vo ze	ked
af be ed ga hy ig ko oc uz vu zi	led
ag bu ce dy eb ik mo ox un xi za	med
aj cy de ef il mu oz ra si uv	ned
ak ca em it ly ob uf yi zu	ped
al en ge if ru uj yd zo	red
am el fo gi iv ke nu oj sa ud yb	sed
an eh fi go ip ka lu od py se um yl	ted
ar cu ez id ma og po ti ux wy	wed
av du ja op re	yed
er je mi no ov ta yu	zed

[redacted]

Terminational Order.

D

eh fi go ip ka lu od py se um yl	bid
ej ix ow ut xa	cid
ap es gy ih ku lo me na of pi uc ym	did
aj cy ef il uv	eid
ag bu ce dy eb xi za	fid
ab bi co da ev gu ij le my on uk ys	gid
at ew ha ki tu yn	hid
ax er iw je mi ov ta up yu	jid
am el fo gi iv ke nu oj sa ud yb	kid
cu ez id ma og po ti ux ve wy	lid
as ek jo or pa te ug wi	mid
ad ec fa hi iq ny ol su vo ze	nid
ex ir ot ry ul	oid
af ed ga hy ig oc uz vu zi	pid
ak ca em it ly ne ob uf yi zu	rid
aw et ic li uq vy wu xe	sid
ci du eq fy iz ja op re ur	tid
al en ge if ji ru yd zo	vid
ac ba do eg fe ib ky om pu ri yx	wid
az ep ju la ni ok us we	zid
ab bi co da ev gu le my on uk ys	ald
ap es gy ku lo na of pi uc ym	eld
el fo gi iv ke nu oj sa ud yb	old
ah ex ir ot pe ry ul wa yo	and
fo gi iv ke nu oj sa ud yb	end
ap es gy ih ku lo me na of uc ym	ind
ab bi co da ev gu ij le my on uk ys	ond
af be ed ga hy ig ko oc uz yt zi	und
ak ca em ly ne ob so uf yi zu	bod
az ep ju la ni ok ro ty we yc	cod
al by en ge if ji os ru uj va yd zo	dod
am el fo gi ke nu oj sa ud yb	fod
an eh fi go ip ka lu od py se um yl	god
cu ma og po ti ux wy	hod
af ed hy ig oc yt	iod
ah ex ir ot pe ul vi wa yo	jod
as ek fu im jo or pa te ug wi	kod
ap es gy ih ku lo me na of uc ym	lod
at ew ha is oy tu uh wo yn	mod
er iw je mi no ov ta yu	nod
av ci du eq fy iz ja op ur	pod
ac ba eg ib ky om pu uw yx	rod
ec iq ny ol ub vo ze	sod
ag ce dy eb un xi za	tod
aw ho ic li uq wu xe ya	vod
ab bi co da ev gu ij le my uk ys	wod
ay di ej ix ow qu ut xa	ard
ac ba do eg fe ib ky om pu ri uw yx	erd
af be ed ga hy ig oc uz vu yt zi	ord
aj cy ef il oz	aud
al en ge if ji ru uj yd zo	bud
er iw je mi no ov sy ta yu	cud
am el fo gi iv ke nu oj sa ud yb	dud
an fi go ip ka lu od py se um yl	fud
at ew ha is ki oy tu uh wo yn	gud
aw et ho ic li uq vy wu xe ya	hud
di ix ow xa ye	jud
ap es gy ih ku lo me na of pi uc ym	kud

D

ab bi co da ev gu ij le my on uk ys	lud
av ci du eq fy iz ja op re ur	mud
az ep ju la ni ok ro ty us we yc	nud
aq	oud
ac ba do eg fe ib ky om pu ri uw yx	pud
as fu im jo pa te wi	rud
af be ed hy ig ko oc uz vu yt zi	sud
he oq xo	tud
ek fu im pa te ug wi	cyd
af be ed ga ig ko oc uz vu zi	hyd
ac ba do eg ib om pu ri uw	kyd
am el fo gi iv ke nu oj sa ud	lyd
an eh fi go ip ka lu od um	myd
ax er iw mi no ov ta up	pyd
ar id ma og po ti ux ve	ryd
al by en ge if os ru va zo	syd
az ep la ni ok ro us we	tyd

E

ak ca em it ly ne ob uf yi zu	abe
am el fo gi iv ke nu oj sa ud yb	ebe
ba do eg ib ky om pu uw yx	ibe
cu ez id ma og po ti ux ve wy	obe
ab bi co ev ij my on uk ys	ube
ed ga hy ig ko uz vu yt	ice
ap gy ih ku lo me na of pi uc ym	oce
ax iw je mi no ov sy ta up yu	ade
ar cu ez id ma og po ux ve wy	ede
bu ce dy eb un za	ide
eh fi ip ka lu od py se um yl	ode
at ew ha is ki oy wo yn	ude
by en ge if os ru uj va yd zo	ife
aq ey oh xu	ofe
ak ca em it ly ne ob so uf yi	ufe
ec iq ny ub vo ze	age
ab bi co da ev gu ij le my on uk ys	ege
ap es gy ih ku lo me na of uc ym	ige
aw et ho ic uq vy wu xe ya	oge
as im jo pa te ug wi	uge
ez id og po	rhe
aq ey oh xu	she
oq xo	the
at ew ha is ki oy tu uh yn	ake
an eh fi go ip ka lu od py se um yl	eke
am el fo gi iv ke nu oj sa ud yb	oke
ay di ej ix ow ut xa ye	ske
an eh go ip ka lu od py se um yl	ale
ec iq ny su ub vo ze	ble
ac ba do fe ib ky om pu ri uw yx	ele
he in oq xo	fle
ax er iw je mi no ov sy ta up yu	gle
et ho uq wu xe ya	ile
ew ha ki oy tu yn	ole
ay ej ix ow qu ut xa ye	ple
aq ey oh xu	tle
ap gy lo of pi uc ym	ule
ak ca em ne ob so uf zu	yle
az ep ju la ni ok ro us we	zle

E

fo gi ke nu oj sa ud	ame
ha is oy tu yn	eme
av du eq fy iz ja to ur	ime
ak ca em it ly ne ob so uf yi zu	ome
ah ex hu ot pe ry ul yo	sme
al by en if ji os uj va yd zo	ume
af be ed ga hy ig ko oc uz vu yt zi	ane
az ep ju la ni ok ro ty us we yc	ene
am fo iv ke nu oj sa ud	ine
ec iq ny ol su ub vo ze	one
an eh fi go ip ka od py se um yl	une
at ew ha is tu wo	yne
as ek fu jo or pa te ug wi	ape
ag bu ce dy eb ik mo ox un xi za	epe
he in oq xo	ipe
az ep ju la ni ok ro ty us we yc	ope
ap es gy ih ku lo of uc ym	are
ax er iw je mi no ov sy ta up yu	bre
am fo gi iv ke nu oj ud yb	cre
aq ey oh xu	dre
aj bo cy de ef il mu ra uv	ere
aw et ho ic li uq vy wu xe ya	fre
at ew ha is ki oy tu uh yn	gre
az ep ju la ok ro ty we yc	ire
di ej ix ow ut xa	ore
ah ex ir ot ry vi wa yo	pre
ab bi co da ev gu le my on uk	tre
ag ce dy eb un xi za	ure
ac ba do eg fe ib ky om pu ri uw yx	ase
ca em it ly ne so uf yi zu	ese
eh go ip ka od py um yl	ise
by en ge if ji os ru uj va yd zo	ose
af ga hy ig ko oc uz yt zi	use
bi co da ij le on uk ys	ate
ap es gy ku lo me na of uc ym	ete
cu ez og po ux ve wy	ite
af be ed ga hy ig uz vu yt zi	ote
az ep la ni ok ty us we yc	ute
av ci du eq fy iz ja op to ur	ave
en ge if ji ru uj yd zo	eve
ax er iw no ov sy ta up yu	ive
ag ce dy eb ik mo ox un xi za	ove
ay di ej ix ut xa ye	uve
aj bo cy de ef il mu oz ra si uv	awe
af ed ga hy ig ko oc uz vu yt zi	ewe
av ci du eq fy iz ja op re	owe
at ew ha ki tu uh wo yn	swe
az ep ju la ni ok ro ty we yc	axe
ad ec fa hi ny ol su ub vo ze	exe
as ek fu im jo or pa te ug wi	oxe
bu ce eb un	aye
ex ot pe wa	eye
er iw ov sy ta	tye
ha is ki oy tu uh	wye
he in oq xo	aze
av ci du eq fy iz ja op re ur	eze
as ek fu im jo or pa te ug	ize
ab bi co da ev gu ij le my on uk ys	oze
ad ec fa hi iq ny ol ub vo ze	uze

Terminational Order.

F

as ek fu im jo or pa te ug wi	baf
ap es gy ih ku lo me na of pi uc ym	caf
ab bi co da ev gu ij le my on uk ys	daf
aw ho ic li uq vy wu xe ya	faf
az ep ju la ni ok ro ty we yc	gaf
ar cu ez id ma og po ve wy	haf
ac ba do eg fe ib om pu ri uw	jaf
ad ec fa hi iq ny ol su ub vo ze	kaf
af ed ga hy ig ko oc uz vu yt zi	laf
ag bu ce dy eb ik mo ox un xi za	maf
aj bo cy de ef il mu oz ra si uv	naf
ah ir ot ry ul	oaf
ak ca em it ly ob so uf zu	paf
al by en ge if os ru uj va yd zo	raf
am el fo gi iv ke nu oj sa ud yb	saf
ay di ej ix ow qu ut xa ye	taf
at ew ha is ki oy tu uh wo yn	waf
av du ja op re to ur	yaf
ax er je mi no ov ta up yu	zaf
ar cu ez id ma og po ti ux ve wy	bef
ah ex hu ir ot pe ry ul vi wa yo	cef
as ek fu im jo or pa te ug wi	def
ab bi co da ev gu ij le my on uk ys	gef
aq ey xu	hef
ax er iw je mi no ov ta up yu	jef
av ci du eq fy iz ja op re to ur	kef
ac ba do eg fe ib ky om pu ri uw yx	lef
ad ec fa hi iq ny su ub vo ze	mef
af be ed ga hy ig ko oc uz vu yt zi	nef
aw et ho vy wu xe ya	pef
aj bo cy de ef il mu oz ra si uv	ref
ak ca em ly ne ob uf zu	sef
al by en ge if ji os ru uj va yd zo	vef
at ew ha is tu wo	yef
ap es ku lo me na of pi	zef
ab bi co da ev gu ij le my on uk ys	bif
al by en ge if ji os ru uj va yd zo	dif
ay di ej ix ow qu ut xa ye	fif
ba do eg fe ib ky om pu ri uw yx	gif
ad ec fa hi iq ny ol su ub	hif
ag bu ce eb ik mo ox un za	jif
ap es gy ih ku lo me na of pi uc ym	kif
aj cy de ef il mu oz ra si uv	lif
ak ca em ly ne ob so uf yi zu	mif
an eh fi go ip lu od py se um yl	nif
ar cu ez ma og po ti ux ve wy	pif
as ek fu im jo pa te ug wi	rif
af be ed ga hy ig ko oc uz vu yt zi	sif
aw et ho ic li uq vy wu xe ya	tif
ax er iw je mi no ov sy ta up yu	vif
he in xo	wif
at ew ha is tu wo	zif
az ep ju la ni ok ro ty we yc	elf
aj bo cy de ef mu oz ra si uv	bof
ak ca em ly ne ob so uf yi zu	cof
av ci du eq fy iz ja op re to ur	dof
am el fo gi iv ke nu oj sa ud yb	gof
af be ed ga hy ig ko oc uz vu yt zi	hof

F

an eh fi go ip ka lu od se um	jof
ay di ej ix ow xa ye	kof
as ek fu pa te ug wi	lof
ap es gy ih ku lo me na of pi ym	mof
at ew ha is ki oy tu wo yn	nof
ah ex hu ir ot ry ul vi wa yo	pof
az ep ju ni ok us we yc	rof
ac ba do eg fe ib ky om pu ri uw yx	sof
ax er iw je mi no ov sy ta up	tof
ag bu ce eb ik mo ox un xi za	zof
am el fo gi iv ke nu oj sa ud yb	erf
an eh fi go ip ka lu od py se um yl	orf
an eh fi go ip ka lu od py se um yl	buf
aw et ho ic li uq wu xe ya	cuf
ar cu ez id ma og po ti ux ve wy	duf
ew ha is ki oy tu uh wo yn	fuf
ap es gy ih ku lo me na of pi uc ym	guf
al en ge if ji ru uj yd zo	huf
ab bi co da ev gu le on uk ys	juf
er iw je mi no ov sy ta yu	kuf
az ep ju la ni ok ro ty us we yc	luf
ac ba do eg fe ib ky om pu ri uw yx	muf
ec iq ny ol su ub vo ze	nuf
af be ed ga hy ig ko oc uz vu yt zi	puf
ag bu ce dy eb ik mo ox un xi za	ruf
aj bo cy de ef il mu oz si uv	suf
am el fo gi iv ke nu oj sa ud yb	tuf
av ci du eq fy iz ja op re to ur	vuf
di ix ow xa ye	zuf
az ep ju la ni ok ro us we	kyf
ak ca em it ne ob so uf	ryf
at ew ha is ki tu wo	syf

G

ar cu ez id ma og po ti ux ve wy	bag
as ek fu jo pa te ug wi	cag
ap es gy ih ku lo na of pi uc ym	dag
bi co da ev gu ij le on uk ys	fag
er iw je mi no ov sy ta yu	gag
ay ej ix ow qu ut xa	hag
ba do eg fe ib ky om pu ri uw yx	kag
ec iq ny ol su ub vo zė	lag
af be ed ga hy ig ko oc uz vu yt zi	mag
ag bu ce dy eb un xi za	nag
aj bo cy de ef il mu oz ra si uv	pag
ak ca em it ly ne ob uf yi zu	rag
al en ge if ru uj yd zo	sag
ah ir ry ul wa yo	tag
am el fo gi iv ke nu oj sa ud yb	vag
av ci du eq fy iz ja op re	wag
an eh fi go ip ka lu	zag
ay	aeg
bi co da ev gu ij le uk ys	beg
al by en ge if os ru uj va yd zo	ceg
an eh fi go ip ka lu od py se um yl	deg
ac ba do eg fe ib ky om pu ri uw yx	feg
ad ec fa hi ny ol su vo ze	geg
ax er iw je mi no ov sy ta up yu	heg

G

af ed ga ko oc vu yt zi	jeg
ag bu ce dy eb un xi za	keg
aj bo cy de ef il mu oz ra si uv	leg
ak ca em it ly ne ob so uf yi zu	meg
ap es gy ih ku lo me na of pi uc ym	neg
am el fo gi iv nu oj sa ud yb	peg
av ci du eq iz ja op re to ur	reg
az ep ju ni ok ro ty we yc	seg
ar cu ez id ma og po ti ux ve wy	teg
as ek fu im jo or pa te ug wi	veg
et ho ic li uq vy wu xe ya	weg
he in xo	zeg
aq ey oh	aig
aj cy de ef il mu oz ra si uv	big
at ew ha is ki oy tu uh wo yn	cig
ar cu ez id og po ti ux ve wy	dig
aw et ic uq	eig
an eh fi go ip lu od py se um yl	fig
ah ex ir ot pe ry ul vi wa yo	gig
ca em it ly ne ob so uf yi zu	hig
as ek fu im or pa te ug wi	jig
az ep ju la ni ok ro ty us we yc	kig
ac ba do ib ky pu ri uw yx	lig
ab bi da ev gu ij le my on uk	mig
al by en ge if ji os ru uj va yd zo	nig
ec iq ny ol su ub vo ze	pig
ag bu dy eb za	rig
el fo gi iv ke nu oj sa ud yb	sig
af be ed ga hy ig oc uz vu yt zi	tig
ay di ej ix ow qu ut xa ye	vig
er iw mi ov sy ta yu	wig
av du iz ja op re to ur	zig
as ek fu im jo or pa te ug wi	ang
ad ec fa ny ol su ub vo ze	eng
ab co da ev gu le my on uk ys	ing
by en ge if ji os ru uj va yd zo	ong
ar ez id ma og po ti uv ve wy	ung
ag bu ce dy eb un xi za	bog
aj bo de ef il oz ra si uv	cog
ak ca em it ly ne ob uf yi zu	dog
az ep ju la ni ok ro ty we yc	fog
al en ge if ji ru uj yd zo	gog
et ho ic uq vy wu xe ya	hog
am el fo gi iv ke nu sa ud yb	jog
an eh fi go ip ka lu od py se um yl	kog
ar cu ez id ma og po ti ux ve wy	log
as ek fu im jo or pa te ug wi	mog
av ci du eq fy iz ja op re ur	nog
ab bi co da ev gu ij le my on uk ys	pog
ah ex hu ot pe ry ul vi wa yo	rog
ap es gy ku lo me na of pi uc ym	sog
ay ej ix ow qu ut xa	tog
ac ba do eg fe ib ky om pu ri uw yx	vog
at ew ha is ki oy tu uh wo yn	wog
ec fa ol su ub vo ze	yog
af ed ga ko vu zi	zog
am el fo gi iv ke nu oj sa ud yb	arg
ak ca em it ly ne ob so uf yi zu	erg

Terminational Order.

G

ad ec fa hi iq ny ol ub vo ze	urg
in oq	aug
af be ed ga hy ig ko uz vu yt zi	bug
ag bu ce dy eb ik mo ox xi za	cug
aj bo cy de ef il mu oz ra uv	dug
ak ca em it ly ne ob uf yi zu	fug
av ci du eq fy iz ja op re to ur	gug
ec iq ny ol su ub vo ze	hug
ac	jug
am el fo gi iv ke nu oj sa ud yb	kug
an eh fi go ip ka lu od py se um	lug
ar cu ez id ma og po ti ux ve wy	mug
as ek fu im jo or pa te ug wi	nug
ep ju la ni ok ro ty we	pug
bi co da ev gu ij le my uk	rug
er iw je no ov sy ta yu	sug
aw et ho ic li uq vy wu xe ya	tug
ap ih ku lo na uc	yug
al en ge if ji os ru va zo	zug
af be ed ga ig ko oc uz vu zi	cyg
al by if ji os ru uj va zo	hyg
ax er iw je mi no ov sy ta up	lyg
ac ba do eg fe ib ky om pu ri uw	myg
an eh fi go ip ka lu od py se um	nyg
he in xo	tyg

H

ay di ej ix ow qu ut xa	bah
at ew ha is ki tu uh wo yn	cah
he in oq xo	dah
as ek fu im jo or pa te ug wi	fah
aq ey ox hu	gah
ax er iw je mi no ov sy ta up yu	hah
ak ca em ne ob so uf zu	jah
aw et ho ic li uq vy wu xe ya	kah
av ci du eq fy iz ja op re to ur	lah
ec fa iq ny ol su ub vo ze	mah
af be ed ga hy ko oc uz vu yt zi	nah
ex hu ot pe ry ul vi wa yo	rah
aj bo cy de ef il mu oz ra si uv	sah
al by en if ji os ru uj va yd zo	tah
ap es gy ih ku lo me na of pi uc	wah
ac ba do eg fe ib om pu	yah
ab bi co da ev gu le on uk	zah
ac ba do eg ib ky om pu ri uw yx	ach
ab bi co da ev gu ij my on uk ys	ech
am el iv ke nu oj sa ud yb	ich
an eh fi go ip ka lu od py se um	och
ay di ej ix ow ut xa	uch
ap gy ih ku lo me na of pi uc ym	feh
ay di ej ix ow ut xa ye	geh
as ek fu im jo or pa te ug wi	heh
ad ec fa hi su ub vo ze	jeh
ax er iw je mi no ov sy ta up yu	keh
am el fo gi iv ke nu oj sa ud yb	leh
an eh fi go ip ka lu od py um yl	meh
ar cu ez id og po ti ux ve wy	neh
az ep ju la ni ok ro ty us we yc	veh
ak ca em it ly ob so uf yi zu	weh
ag bu ce eb ik mo ox un xi za	zeh
ad ec fa hi iq ny ol su ub vo ze	boh
av ci du eq fy iz ja op re to ur	foh
ak ca em it ly ne ob so uf yi zu	loh
ab bi co da ev gu ij my on uk ys	moh
he in oq xo	poh
at ew ha is ki oy tu uh wo	roh
eh fi go ip ka lu od py se um yl	aph
ac ba do eg fe ib ky om pu ri uw yx	eph
aj bo cy ef il mu oz ra uv	iph
ay di ej ix ow ut xa	oph
al by en if ji os uj va yd zo	uph
al en ge ru uj yd zo	ash
at ew ha is oy tu uh wo yn	esh
ax er iw je mi no ov sy ta up	ush
ag bu ce dy eb ik mo ox un xi za	ath
eh fi ka lu od py se um yl	eth
ab da ev gu my on uk ys	ith
ap gy ku lo me na of pi uc ym	oth
ah ir	puh
aw et ho ic li wu xe ya	suh

I

ad fa hi iq ny su vo ze	abi
at ew ha is ki oy tu uh wo yn	ebi
ab co da ev gu le my on uk ys	ibi
ag bu ce dy eb un xi za	obi
as ek im jo or pa ug wi	ubi
al by en ge if ji os ru uj va yd zo	aci
an eh go ip ka lu od py se um	ici
be ed ga hy ig ko oc uz vu zi	oci
ad ec fa hi iq ny ol su ub vo ze	edi
cu ez ma og po ux wy	idi
aj bo cy de ef il mu oz ra si uv	odi
ap es gy ih lo me na of pi uc ym	udi
af be ed ga hy ig ko oc uz vu zi	afi
an eh fi go ip ka lu od py se um yl	efi
az ep ju la ok ty us we yc	ifi
ac ba do eg ib ky om pu ri uw yx	ofi
at ew ha is ki oy uh wo yn	ufi
ak ca em it ly ne ob so uf yi zu	agi
ag bu ce dy eb ik mo ox un xi za	egi
ec fa iq ny ol ub vo ze	igi
ax iw je mi no ov sy ta up yu	ogi
az ep la ni ok ty us we yc	ugi
ej ow ut xa	ahi
ap gy ku lo me na of pi	chi
ad ec ny ol su ub vo	rhi
in oq xo	shi
eg fe ib ky om ri uw yx	thi
aj bo cy de ef il ra si uv	aki
ab co da ev gu ij le on uk ys	eki
av du eq fy iz ja op re to ur	oki
ag bu ce dy eb ik mo ox un xi za	ski
ez id ma og po ti ux ve wy	uki
af be ed ga hy ig ko oc uz vu yt zi	bli
am el fo gi iv ke nu oj sa ud yb	cli
by en ge if ji os ru uj va yd zo	eli
aw et ho ic li uq vy wu xe ya	fli
at ew ha is ki oy tu uh wo yn	gli
bu dy eb ik mo ox un za	ili
ad ec fa hi iq ny ol su ub vo ze	kli
ap es gy ih ku lo me na of pi uc ym	oli
as ek fu or pa ug wi	pli
di ej ix ow xa ye	sli
ax er iw je mi no ov sy ta up	uli
ce dy eb un xi za	ami
af be ga hy ig ko oc uz vu yt zi	emi
by en ge if os ru uj va yd zo	imi
an eh fi go ip ka lu od py se um yl	omi
cu ez id ma og po ti ux ve wy	smi
ba do eg fe ib ky om ri uw yx	umi
an eh fi go ip ka lu od py se yl	ani
ex hu ir ot pe ry ul wa yo	eni
at ew ha is oy tu uh wo yn	ini
ar cu ez id ma og po ti ux ve wy	oni
ak ca em it ly ne ob so uf yi	uni
az ep ju la ni ok ro ty us we yc	api
ap es gy ih ku lo na of pi uc ym	epi
af be ed ga hy ig ko uz vu yt	ipi
as ek fu im jo or pa te ug	opi
ax er mi no ov sy ta up yu	spi
aw et ho ic li uq xe ya	upi
ar cu ez id ma og po ti ux ve wy	ari
he in oq xo	bri
bi co da ev gu ij le uk ys	cri
at ew ha ki oy uh wo	dri
am el fo gi iv ke nu oj sa ud yb	fri
av du eq fy iz ja op to ur	iri
aw et ho ic li uq wu xe ya	kri
ak ca it so uf yi zu	ori
aq ey oh xu	pri
di ej ix ow xa ye	uri
ap es gy ih ku lo me na of uc ym	asi
er iw je mi no ov sy ta yu	esi
fo iv ke nu oj ud yb	isi
ec fa hi iq ny ol su ub vo ze	osi
ex ot ry ul vi wa yo	usi
ar cu id ma og ti ux ve	eti
ak ca em it ly ne zu	iti
az ep ju la ok ro ty we	oti
an ip ka lu py um	sti
af be ed ga hy ig ko oc uz yt zi	uti
at ew ha is ki oy tu uh wo yn	avi
el fo gi iv oj sa ud yb	ovi
al by en ge if ji uj va yd zo	uvi
ac ba do eg fe ib ky om pu ri	ewi
ab bi co da ev gu ij le my on uk ys	axi
av ci du eq fy iz ja op re to ur	exi
ay ix ut xa	ixi
at ew ha ki tu uh wo yn	oxi
av ci du eq fy iz ja op re to ur	azi
as ek fu im or pa te ug wi	ezi
ax er iw no ov ta up yu	izi
ay di ej ix ow ut xa ye	ozi
am el fo gi iv ke oj sa ud yb	uzi

Terminational Order.

J

az ep la ni ok ro ty us we	baj
ay di ix ow xa	caj
bo cy de ef il mu oz ra si uv	haj
eh fi go ip ka lu od py se um yl	maj
at ew ha is ki oy tu uh wo	paj
aw et ho ic li wu xe ya	raj
ag bu ce dy eb ik mo ox un xi za	dej
ap es ih ku lo me na pi uc	vej
ad ec fa hi ny ol su ub vo ze	bij
al by en ge if os ru va zo	pij
af be ed ga hy ig oc	rij
az ep la ni ok ro ty us we	tij
ag bu ce dy eb ik mo ox un za	vij
aw et ho ic li uq vy wu xe ya	boj
ay di ix ow ut ye	roj

K

an eh fi go ip lu od py se um yl	bak
ar cu ez id ma og po ti ux ve wy	dak
he in oq xo	fak
ap es gy ih ku lo me na of pi uc yn	gak
el fo gi iv ke nu oj sa ud yb	hak
ab bi co da ev gu le on uk ys	jak
ax er iw je mi no ov sy ta up yu	kak
ay di ej ix ow ut xa ye	lak
ac ba do eg fe ib ky om pu ri uw	mak
at ew ha is ki oy tu uh wo yn	nak
ec iq ny ol su ub vo ze	oak
af be ed ga hy ig ko oc uz vu yt zi	pak
ag bu ce dy eb ik mo ox un za	rak
ah ex hu ir ot pe ry ul vi wa yo	sak
az ep ju la ni ok ro ty us we yc	tak
ak ca em it ly ne ob so uf yi zu	vak
as ek fu im jo or pa te ug wi	wak
al ge if ru zo	yak
aw ho ic li wu xe ya	zak
ak ca em it ly ne ob uf yi zu	ack
az ep ju la ok ro ty we yc	ick
ar cu ez id ma og po ti ux ve wy	ock
el fo gi iv ke oj sa ud yb	uck
af be ed ga hy ig ko uz vu yt zi	bek
aj bo cy de ef il mu oz ra uv	dek
av ci du eq fy iz ja op re ur	gek
ak ca em it ly ob so uf yi	hek
at ew ha is ki tu uh wo	jek
am el fo gi iv ke nu oj sa ud yb	kek
an eh fi go ip ka lu od py se um	lek
ar cu ez id ma og po ti ux ve wy	mek
ac ba do eg fe ib ky om pu ri uw yx	nek
az ep ju la ni ok ro ty we	pek
ab bi co da ev gu ij le my uk	rek
ax er iw je no ov sy ta up yu	sek
ah ex hu ir ot pe ry ul vi wa yo	tek
ec fa hi ol su ub vo ze	zek
am el iv ud yb	aik
ap es gy ih of pi uc ym	bik
er iw je ov sy ta yu	dik
az ep ju la ni ok ro ty us yc	fik

K

ag bu ce dy eb un xi za	hik
ar cu id ma og po ux ve	jik
ay ej ix ow qu ut xa ye	kik
as fu im jo pa te	lik
an eh go ip ka lu od py se um yl	mik
ak em it ly ne ob uf yi zu	nik
ab bi co da ev gu ij my on uk ys	pik
ba eg fe ib ky om ri uw yx	rik
al en ge if ru uj va yd zo	sik
ad ec iq ny ol su vo ze	tik
af be ed ga hy ig ko oc uz vu yt zi	vik
de ef il mu oz ra uv	zik
al en ge if ji ru uj yd zo	alk
at ew ha is ki oy tu uh yn	elk
ar cu ez id ma og po ux ve wy	ilk
av ci eq fy iz ja op re to ur	ulk
az ep ju la ni ok ro ty us we yc	ank
aw et ho ic uq wu xe ya	ink
ay di ej ix ow ut xa ye	unk
ax er iw je mi no ov sy ta up yu	bok
ad ec fa hi iq ny ol su ub vo ze	dok
af be ed ga hy ig ko oc uz vu yt zi	fok
ab co da ev gu ij le my on	hok
aw et ho ic li wu xe ya	jok
ak ca em it ly ne ob so uf yi zu	kok
at ew ha is ki oy tu uh yn	lok
al by en if ji os ru uj va yd zo	mok
am el fo gi iv ke nu sa ud yb	nok
an eh fi go ip ka lu od py se um	pok
as ek fu im jo or pa te ug wi	rok
aq ey oh xu	sok
he in oq xo	tok
av ci du eq fy iz ja op re to ur	vok
ix ow ut xa	yok
aw et ho ic li uq vy wu xe ya	ark
ay di ej ix ow ut xa ye	erk
ap gy ih ku lo me na of uc ym	irk
ax er iw je mi no ov sy ta up yu	ork
ag bu ce dy eb ik mo ox un xi za	ask
at ew ha tu uh wo yn	isk
aw et ho uq vy wu xe ya	osk
ev gu ij on uk ys	auk
av ci du eq fy iz ja op re to ur	buk
ac ba do eg fe ib ky om pu ri uw yx	duk
aj bo cy de ef il mu oz ra si uv	huk
ag bu ce eb un xi za	juk
at ew ha is ki oy tu uh wo yn	kuk
ah ex ir ot pe ry ul vi wa yo	luk
ax er iw je mi no ov sy ta up yu	muk
al by en ge if ji os ru va yd zo	nuk
ar cu ez id ma og ti ux ve wy	puk
an eh fi go ip ka lu od py se um yl	ruk
ep ju la ni ok ro us we	suk
ak ca em it ly ne ob uf zu	tuk
am fo gi iv ke nu sa ud	zuk
af be ed ga hy ig ko oc uz vu yt zi	awk
ap es gy ih ku lo na of pi uc	cyk
ab bi co da ev gu ij le my on uk	kyk

K

av ci du fy iz ja op re to ur	nyk
ek fu im jo or pa te ug wi	pyk
af be ed ga hy ig ko oc uz vu zi	syk

L

am el fo gi iv ke nu oj ud yb	bal
ah ex ir ot pe ry ul wa yo	cal
er iw je mi ov sy ta yu	dal
aw et uq	eal
ay di ej ix ow qu ut xa ye	fal
as ek fu jo pa te ug wi	gal
aq ey	ial
ap es ih ku lo me na of pi uc	jal
ab bi co da ev gu ij le my on uk ys	kal
ak ca em it ly ne ob so uf yi zu	lal
az ep ju la ni ok ty we yc	mal
ac do eg fe ib ky om ri uw yx	nal
ec iq ny ol su ub vo	pal
af be ed ga hy ig ko oc uz vu yt zi	ral
ag bu ce dy eb un xi za	sal
cu ez id ma og po ti ux ve wy	tal
at ew is oy uh yn	ual
aj bo cy de ef il mu oz ra si uv	val
an fi go ip ka lu od py se um yl	wal
av ci du fy iz ja op re to ur	xal
al ge if ji ru va zo	zal
al by en ge if ji uj va yd zo	bel
ax er iw je mi no ov sy ta up yu	cel
am el gi iv ke oj sa ud yb	del
an fi go ip ka lu od py se um yl	fel
ar cu id ma og po ti ux ve wy	gel
as ek fu im jo or te ug wi	jel
ap es gy ih ku lo me na of pi uc ym	kel
ab bi co da ev gu ij le my on uk ys	lel
av ci du eq fy iz ja op re to ur	mel
az ep ju la ni ok ro ty us we yc	nel
ac ba do eg ib ky pu ri uw yx	pel
ay di ej ix ow qu ut xa	rel
af be ed hy ig ko oc uz vu yt	sel
aw et ic li uq vy wu xe ya	tel
ey oh	uel
ag bu ce dy eb ik mo ox un xi	vel
at ew ha ki tu uh wo yn	wel
ad ec fa hi ny ol su ub vo ze	xel
ex hu ir ot pe ul wa	yel
ak ca em it ob so uf zu	zel
az ep ju la ok ro ty us we yc	bil
as ek fu im jo pa te ug wi	cil
ab bi co da ev gu ij le my on uk ys	dil
av ci du eq fy iz ja op re ur	fil
ag bu ce dy eb ik mo ox un xi za	gil
ap gy ku lo na of pi ym	hil
an eh fi go ip ka lu od se um	jil
ac ba do eg fe ib ky om pu ri uw yx	kil
ah ex hu ir pe ry wa yo	lil
af be ed ga hy ig ko oc uz vu yt zi	mil
aw et ho ic li uq vy wu xe ya	nil
ca it ly ne ob so uf yi zu	pil

[redacted]

Terminational Order.

L

al by en ge if os uj va yd	ril
ec iq ny ol su ub vo ze	sil
ax er iw je no sy ta up yu	til
am fo gi iv ke nu oj sa ud yb	vil
ay di ej ix ow qu ut xa	wil
ap es gy ih ku lo me na of pi uc ym	bol
bi co da gu ij le on uk ys	col
ah ex ir ot pe ry ul vi wa yo	dol
at ew ha is ki oy tu uh wo	fol
ac ba do eg fe ib ky om pu uw yx	gol
ax mi ov sy ta up yu	hol
ix ow ut xa ye	jol
af be ed ga hy ig ko oc uz vu yt zi	kol
ag bu ce dy eb ik mo ox un xi za	lol
aj de ef il mu oz ra uv	mol
ak em it ly ne ob so uf yi zu	nol
al by en ge if ji os ru uj va yd zo	pol
am el fo iv ke nu oj sa ud yb	rol
an go ka od py se um yl	sol
ec ny ol su ub vo ze	tol
ar cu ez id ma og po ti ux ve wy	vol
as ek fu jo pa te	wol
az ep ju la ni ok ro ty us we	xol
aw ho ic wu	yol
av ci du iz ja op re to ur	zol
ad ec fa hi iq ny ol su ub vo ze	erl
af be ed ga hy ig ko oc uz vu yt	irl
ap es gy ku lo me na of uc	isl
as ek im ug	aul
ah ex hu ir ot pe ry ul vi wa yo	bul
at ew ha is oy tu uh wo	cul
an eh fi go ip lu od py se um yl	dul
al by en ge ji os ru uj va yd zo	ful
ab bi co da ev gu ij le my on uk	gul
ac ba do eg fe ib ky om pu ri uw yx	hul
ep la ni ok us we	jul
ag bu ce dy eb un xi za	kul
ax er je mi no ov sy ta up yu	lul
et ho ic li uq vy xe ya	mul
ay di ej ix ow qu ut xa ye	nul
he oq xo	pul
ap es gy ku lo me na of pi ym	rul
av ci du eq fy iz ja op re to ur	sul
al bo cy de ef il mu oz ra si uv	tul
ak ca em it ly ne ob so uf yi zu	vul
af be ed ga ko vu zi	zul
ap es gy ih ku lo na of pi uc ym	awl
ac ba do eg ib ky om pu ri uw yx	owl
ar cu ez id ma og po ti ux ve wy	byl
at ew ha is ki oy tu uh wo	dyl
as ek fu jo or pa te ug wi	fyl
ad ec fa ny ol su ub vo ze	gyl
ah ex hu ir pe ul vi wa	hyl
aw et ho ic li wu xe	lyl
ab bi co da ev gu ij le my on uk	myl
af be ed ga hy ig ko oc uz zi	nyl
ba eg fe ib ky om ri uw	ryl
ax er iw je mi no ov sy ta	syl
ag dy eb ik mo ox un xi za	tyl

M

at is oy yn	sam
en ge if ji ru uj yd zo	bam
er iw je mi no ov sy ta yu	cam
am el fo gi iv ke nu oj sa ud yb	dam
an fi go ip ka lu od py se um yl	fam
ar cu id ma og po ti ux ve wy	gam
aw ho ic li uq vy wu xe ya	ham
in xo	jam
ap es gy ih ku lo na of pi uc ym	kam
ab bi co da ev ij le my on uk ys	lam
av ci du eq fy iz ja op re ur	mam
az ep ju la ni ok ro ty we yc	nam
ac ba do eg ib ky pu uw yx	pam
ay di ej ix ow qu ut xa	ram
af be ed hy ig ko oc uz vu yt zi	sam
aj cy de ef il mu oz si	tam
ag bu dy eb ik mo ox un xi za	vam
ah ex hu ot pe ry vi wa yo	wam
ak ca em it ob so uf zu	zam
an eh fi go ip ka lu od py um yl	bem
av ci eq fy ja ur	cem
ar cu ez id ma og po ti ux ve wy	dem
as ek fu im jo or pa te ug wi	fem
ap es gy ih ku lo na of pi uc ym	gem
ay di ej ix ow qu xa	hem
ab bi co da ev gu le on uk ys	jem
at ew is ki oy tu uh wo yn	kem
az ep ju la ni ok ro ty us we yc	lem
ac ba do eg fe ib ky om pu ri uw yx	mem
ad ec hi iq ny ol su ub vo zo	nem
af be ed ga hy ig ko oc vu yt zi	pem
ag bu ce dy eb ik mo ox un za	rem
aj bo cy de ef il mu oz ra si uv	sem
am el fo gi iv ke nu oj sa ud yb	tem
ak em it ly ne so uf yi zu	vem
al en ge if ji ru uj yd zo	wem
ah ir ot ry ul vi wa yo	xem
er je no ov ta up	yem
aw et ho ic li wu xe ya	zem
av ci du fy iz ja op re ur	ohm
ac do ib ky om uw yx	aim
ag ce dy ik ox xi za	bim
ex hu ir ot pe ry ul wa yo	cim
as ek fu ug wi	dim
ap es gy ih me na pi uc ym	fim
ak ca em it ly ne ob so uf yi zu	gim
aq xu	him
ec ol su ub vo ze	jim
cu ez id ma og po ti ux ve wy	kim
eh go ip ka lu od py	lim
ax er iw je ov sy up yu	mim
am el gi iv oj sa ud yb	nim
aj bo cy de ef il mu oz uv	pim
in oq xo	rim
da ev gu ij le my on uk ys	sim
by if ji uj yd zo	tim
av ci du eq fy iz ja op re ur	vim
et ic li uq vy wu	wim

M

at ew ha is ki oy tu uh	xim
ix ow xa	zim
af be ed ga hy ig ko oc uz vu yt zi	alm
an eh fi go ip ka lu od py se um yl	elm
aw et ho ic li uq vy wu xe ya	olm
in oq xo	bom
al en ge if ji os ru va yd zo	com
at ew ha is ki oy tu uh yn	dom
ac ba do eg fe ib ky om pu ri uw yx	fom
ad ec fa hi ny ol su vo ze	gom
aq ey oh xu	kom
aj bo cy de ef il mu oz ra si uv	lom
ak ca em it ly ne ob so uf yi zu	mom
ap es gy ih ku lo of pi uc ym	nom
am el fo gi iv nu oj sa ud yb	pom
av du eq iz ja op re to ur	rom
az ju la ni ok ro ty we yc	som
aw et ho ic li uq vy wu xe ya	tom
as ek fu im jo or pa te ug wi	vom
ax er iw je mi no ov sy ta up yu	wom
an go ip ka lu od se um	yom
ez id ma og po ux	zom
bi co da ev gu ij le on uk ys	arm
ar cu ez id ma og po ti ux ve wy	asm
ec iq ny ol su ub vo ze	ism
aq oh xu	aum
ak ca em it ly ne ob so uf zu	bum
az ep ju la ni ok ro ty we yc	cum
aw et ho ic li uq vy wu xe ya	dum
in oq xo	eum
am el fo gi ke nu oj sa ud yb	fum
an eh fi go ip ka lu od py se um yl	gum
ag bu ce dy eb un xi za	hum
ah ex hu ir ot pe ul vi wa yo	jum
as ek fu im jo or pa te ug wi	kum
ap es gy ih ku lo na of uc ym	lum
bi da ev gu ij le on uk ys	mum
ax er iw je no ov sy up yu	num
av ci du eq fy iz ja op re ur	pum
ac ba eg ib ky om yx	rum
ec iq ny ol ub vo ze	sum
ar cu ez id ma og ti ux wy	tum
af be ed ga hy ig ko oc uz vu yt zi	vum
at ew ha is tu wo	yum
bo de ef il mu oz ra si uv	zum
at ew ha is oy tu wo	cym
ah ex hu ir ot pe ry ul wa	dym
am el fo iv ke nu sa ud	gym
an fi go ip ka lu od py se	hym
ak ca em ly ne ob so uf zu	kym
ar cu ez id ma og po ti ux ve	lym
aq oh xu	mym
ap es ih ku lo me na of pi uc	rym
as ek fu im jo or pa te ug	sym
ay di ej ix ow ut xa	tym

Terminational Order.

N

ak em it ly ne ob uf yi zu	ban
az ep ju la ni ok ro ty	can
al by ge if ji os ru va yd zo	dan
ej ix ow ut	ean
am el fo gi iv ke nu oj sa ud yb	fan
an eh fi go ip ka lu od py se um yl	gan
ah ex hu ir ot pe ry ul vi wa yo	han
ar cu id ma og po ti ux ve	jan
as ek fu jo pa te ug wi	kan
ap es gy ih ku lo me na of pi uc ym	lan
bi ev gu on uk ys	man
ax er iw je mi no ov sy ta up yu	nan
av ci du eq fy iz op re ur	pan
ac ba eg ib ky om pu yx	ran
at ew ha is ki oy tu uh yn	tan
af be ed ga hy ig ko oc uz vu yt zi	van
ag bu ce dy eb un xi za	wan
aw ho ic li uq wu ya	xan
bo de ef il mu oz ra si uv	zan
am fo iv ke nu oj sa ud yb	ben
aw et ho ic li uq wu xe ya	cen
iw mi no ov sy ta yu	den
cu ez id ma og po ti ux ve wy	fen
as ek im jo or pa te ug wi	gen
ap es ih ku lo na of pi uc	jen
bi co da ev gu le on uk ys	ken
av ci du eq fy iz ja op re to ur	len
az ep ju la ni ok ty yc	men
he oq xo	nen
ah ex ir ot ry ul	oen
ec iq ny ol su ub vo	pen
af be ed ga hy ig oc uz yt zi	ren
ag ce dy eb un xi za	sen
ca em ly ne ob uf yi zu	ten
aj cy de ef il mu oz si uv	ven
ay di ej ix ow qu ut xa	wen
at ew ha is ki oy tu	xen
ac ba eg fe ib om pu	yen
al ge if ji ru va zo	zen
ew ha is ki oy tu uh wo	ogn
aq ey oh xu	ohn
aw et ic uq vy	ain
iw je mi ov sy ta yu	bin
ag bu ce dy eb ik mo ox un za	cin
an eh fi go ip ka lu py se um yl	din
be ed ga hy ig ko oc uz vu yt zi	fin
ad ec fa hi ny ol su vo ze	gin
as ek fu jo pa te ug wi	hin
ix ow ut xa	jin
ak ca em it ly ne ob uf yi zu	kin
az ep ju la ni ok ro ty we yc	lin
al by en ge if os ru uj va yd zo	min
ah ir ot pe ry ul vi wa yo	nin
am el fo gi iv ke nu oj ud yb	pin
aq oh xu	rin
es ih ku na of pi uc ym	sin
ar cu ez id og ti ux ve wy	tin
ac do eg ib ky om pu ri uw yx	vin

N

av ci du eq fy iz ja op ur	win
bo ef il mu oz ra si uv	xin
ab bi co da ev gu le on uk	zin
aq oh	aon
av ci du eq fy iz op re ur	bon
am el fo iv nu oj sa ud yb	con
ac ba do eg ib ky om pu ri uw yx	don
ad ec fa hi iq ny ol su ub vo	fon
af be ed ga hy ig ko oc uz vu yt zi	gon
ca em it ly ne ob so uf yi zu	hon
ah ry ul	ion
ag bu ce eb ik mo ox un xi za	jon
aj bo cy de ef il mu oz ra si uv	kon
bi da gu ij le my on uk ys	lon
ax er iw je mi no ov sy ta up yu	mon
al en ge if ji ru zo	non
ez id ma og ti ux ve wy	pon
an eh go ip ka od py um yl	ron
as ek fu jo or pa ug wi	son
ew ha is ki oy tu uh yn	von
az ep ju la ni ok ro ty us we yc	won
ap ku lo na uc	yon
di ix ow xa ye	zon
ak ca em it ly ne ob so uf yi zu	arn
ar cu ez id ma og po ti ux ve wy	ern
aw et ho ic li uq vy wu xe ya	orn
ba do eg ib ky om ri uw yx	urn
ar cu ez id ma og po ti ux ve wy	bun
as ek fu jo pa te ug wi	cun
ap es gy ih ku lo na of pi uc ym	dun
bi co da ev gu ij le on uk ys	fun
er iw je mi no ov sy ta yu	gun
an fi go ip ka lu py yl	hun
aw et ho ic li wu xe ya	jun
ac ba do eg fe ib ky om pu ri uw yx	kun
ad ec hi iq ny ol su ub vo ze	lun
af be ed ga hy ig ko oc uz vu yt zi	mun
ag bu ce dy eb xi za	nun
aj bo cy de ef il mu oz ra si uv	pun
ak ca em it ly ne ob uf yi zu	run
al en ge if ru uj yd zo	sun
az ep ju la ni ok ro ty we yc	tun
am el fo gi iv ke nu oj sa ud yb	vun
he xo	yun
oh xu	zun
bi co da ev gu ij le on uk ys	awn
af ed ga hy ig oc uz vu yt zi	own
ac ba do eg ib ky om pu ri uw	byn
ab bi co da ev gu ij le on uk	cyn
ar cu ez id ma og po ti ux	dyn
ak ca em it ne ob uf zu	gyn
ag bu ce dy eb ik mo ox un xi za	kyn
af be ed ga hy ig ko oc uz vu zi	lyn
am fo gi iv ke nu oj sa ud	myn
ad ec fa hi iq ny ol su ub ze	ryn
az ep ju la ni ok ro ty us we	syn
al by en ge if ji os ru va	tyn
di ej ix ow ut xa	vyn

N

er iw je mi no ov ta	wyn
ah hu ir ul vi wa	xyn

O

ab bi co da ev gu ij le my on uk ys	abo
ar cu ez id ma og po ti ve wy	ebo
aw et ho ic vy wu xe ya	ibo
ak ca em it ly ne ob so uf yi zu	obo
an eh fi go ip ka od py se um yl	ubo
am el fo iv nu oj sa ud	aco
aw et ho ic li uq wu xe ya	eco
ax er iw je no sy ta up yu	ico
av ci du eq fy iz ja op re to ur	oco
by en ge if ji os uj va yd zo	uco
en ge if ji uj yd zo	ado
bi co da ev gu ij le my on uk ys	edo
fo iv ke nu oj sa ud yb	ido
es gy ih ku lo me na of pi uc ym	odo
ce dy eb un xi za	udo
ek fu jo pa te ug wi	afo
ak ca em it ly ne ob so uf yi zu	efo
ac ba do eg fe ib ky om pu uw yx	ifo
am el fo gi iv ke nu oj sa ud yb	ofo
be ed ga hy ig ko oc uz yt zi	ufo
be ed ga hy ig ko oc uz vu yt zi	ago
al en ge if ji uj yd zo	ego
aj bo cy de ef il mu oz ra uv	igo
ag bu ce dy eb ik mo ox un xi za	ogo
ax er iw je mi no ov sy ta up	ugo
at ew ha is tu	ygo
cu ez ma og po ti ux wy	aho
hc oq xo	cho
aj cy de ef mu ra si uv	oho
at ew ha is oy tu uh wo	pho
af be ed hy ig ko oc uz vu yt zi	sho
ah ex hu ir ot pe ry ul vi wa yo	tho
ap es gy ih ku lo na pi uc	who
ew ha is ki oy tu uh wo yn	ajo
ep la ni ok ro ty us yc	ejo
ba do eg fe ib ky om pu ri uw yx	ojo
av ci eq fy iz ja op re to	ujo
he in oq xo	ako
as ek fu im jo or pa te ug	eko
ab co da ev gu le my on uk	iko
az ep ju la ni ok ro ty us we yc	oko
at ew ha is ki oy tu uh wo yn	sko
ap es gy ih lo me na of pi uc ym	uko
di ix ow qu ut xa ye	alo
ac ba eg fe ib ky om pu ri uw yx	blo
aw et ho ic li uq vy wu	clo
aq ey oh xu	elo
ax er iw je mi no ov sy ta up yu	flo
he in xo	glo
an go ip ka lu od py se um	ilo
ah ex hu ir ot pe ry vi wy yo	klo
ar cu ez id ma og ti ux ve wy	olo
af be ed ga hy ig ko oc uz vu yt zi	plo
as ek im jo or te ug wi	ulo

[redacted]

Terminational Order.

O

iw je mi no sy ta yu .... amo
ec fa iq ny ub vo ze .... emo
ak ca ly ne so uf zu................ imo
by en ge if ji os ru uj va yd zo....... omo
aw et ic li uq vy xe ya .... umo
ap es gy ih ku lo me na of uc ym .... ano
ce dy eb ik mo ox un xi za .... eno
az ep ju ni ok ro ty us we .... gno
ek fu im jo or pa te ug .... ino
ax er iw je mi no ov sy ta up yu ..... ono
am el fo gi iv ke oj sa ud yb .... uno
ad ec fa hi iq ny ol su ub vo ze ...... apo
av ci du eq fy ja op re to ur .... epo
ag bu dy eb ik mo ox za............ ipo
aw et ho ic li uq vy wu xe ya .... opo
az ep la ni ok ro ty us we yc .... upo
ci du fy iz ja op to ur............... aro
ab bi da ev gu ij le my on uk ys...... bro
ac ba do eg fe ib ky om pu ri uw yx .. cro
am el fo gi iv ke nu oj sa ud yb ...... dro
be ed ga ig ko oc vu yt zi .... ero
al by en ge if ji os ru uj yd zo .... fro
ah ex hu ir ot pe ry ul vi wa yo...... gro
ad fa iq ny ol ub vo .... iro
ap es gy ih ku lo me na of pi uc ym .. kro
ew ha is ki oy tu uh wo yn .... oro
iw je mi no ov sy ta up yu .... pro
aq ey oh xu .... tro
bo cy ef mu ra uv .... eso
by ge if ru uj va yd .... iso
ej ix ow ut xa ye .... oso
ar ez id ma og po ti ux ve wy........ uso
eh ip ka lu od py um yl .... ato
ba eg ky om ri uw................. eto
he oq xo .... ito
bi co da ev gu ij my uk ys .... oto
aw ho ic uq vy wu ya .... sto
ay di ej ix ow xa ye................ uto
az ep ju la ni ok ro ty us we yc ...... avo
di ix ow qu ut xa .... evo
ap es gy ih ku lo me na of uc ym .... ivo
as ek fu im jo or pa te ug wi .... ovo
ak ca em it ly ne ob so uf yi zu ...... awo
am el fo gi iv ke nu oj sa ud yb...... ewo
ec fa hi iq ny ol su ub vo ze .... owo
ag bu ce dy eb ik mo ox un xi za..... axo
hu ir ot pe ry vi wa yo .... exo
av du eq fy iz ja op re to ur .... ixo
af be ed ga hy ig ko oc uz vu yt zi ... oxo
ec fa hi iq ny ol ub vo .... uxo
aw et ho ic li wu xe .... ayo
ac ba do eg fe ib ky om pu ri .... nyo
ac ba do eg ib ky om pu ri uw yx .... azo
he in oq xo....................... ezo
ew ha oy tu uh wo yn .... izo
ah ex hu ir ot pe ry ul vi wa......... ozo
ca it ne ob so yi .... uzo

P

aj bo cy de ef il mu oz ra si uv ..... bap
ak ca em it ly ne ob uf yi zu......... cap
av ci du eq fy iz ja op re ur .... dap
al en ge if ji ru uj yd zo .... fap
am el fo iv ke nu oj sa ud yb .... gap
af be ed ga hy ig oc uz vu yt zi ...... hap
aw et ic uq .... iap
an eh fi go ip ka lu od se um .... jap
ar cu ez id ma og po ti ux ve wy..... kap
as ek fu jo or pa te ug wi .... lap
ap es gy ih ku lo na of pi uc ym ..... map
bi co da ev gu ij le on uk ys......... nap
iw je mi no ov sy ta yu .... pap
az ep ju la ni ok ro ty we .... rap
ac ba do eg ib ky om pu uw yx ...... sap
ag bu ce dy eb un xi za .... tap
ec iq ny ol su ub vo ze............. vap
iu oq xo .... wap
ah ex ir ot pe ul wa .... yap
at ew ha ki tu wo .... zap
ah ex ot ul...................... aep
ac ba do eg ib ky om pu ri uw yx .... cep
ad ec fa hi iq ny ol su ub vo ze ...... dep
af be ed ga hy ig ko oc uz vu yt zi ... fep
ag bu ce dy eb ik mo ox un xi za..... gep
ak ca em it ly ne ob so uf yi zu ...... kep
at ew ha is ki oy tu uh wo yn........ lep
al en ge if ji ru uj yd zo............ mep
am el fo gi iv ke nu oj sa ud yb ..... nep
an eh fi go ip ka lu od py se um yl... pep
ek fu jo or pa te ug wi .... rep
cu ez id ma og po ti ux ve wy .... sep
aj bo cy de ef il mu oz ra si uv ...... tep
aw et ho ic li uq wu xe ya........... vep
aq ey oh xu .... wep
di ix ow ut xa .... zep
he in oq xo .... cip
ak ca em it ly ne ob uf yi zu......... dip
ad ec fa hi iq ny ol su ub vo ze ...... fip
ar cu ez id ma og po ti ux ve wy..... gip
am el fo gi iv nu oj sa ud yb .... hip
aw et ho ic wu ya .... jip
as ek fu jo or pa te ug wi .... kip
af be ga hy ig ko oc uz vu zi......... lip
ac ba do eg fe ib ky om pu ri uw yx.. mip
ci du eq fy iz ja re ur .... nip
ah ex ir ot pe ry ul vi wa .... pip
gy ih ku lo na of pi uc ym .... rip
aj cy de ef il mu oz ra si uv .... sip
an eh fi go ip ka lu od py se um yl... tip
az ep ju la ni ok ro ty us we yc ...... vip
ew ha is ki oy tu uh wo............ wip
al ge if ji ru zo .... zip
aw et ho ic li uq vy wu xe ya........ alp
ay di ej ix ow ut xa ye.............. amp
ab bi co da ev gu ij le my uk ys...... emp
ac ba do eg ib ky om pu uw yx ...... imp
ec fa hi iq ny su ub vo ze .... omp

P

af be ed ga hy ig ko oc uz vu zi...... cop
ag bu ce dy eb ik mo ox un xi za..... dop
aj bo cy de ef il mu oz ra si uv ...... fop
ak ca em it ly ne ob so uf yi zu ...... gop
ix ow qu ut xa.................... hop
he in xo .... jop
ah ex ir ot pe ry ul vi wa yo......... lop
an eh fi go ip ka lu od py se um yl... mop
ar cu id ma og po ti ux ve wy .... nop
as ek fu jo or pa ug wi .... pop
bi da ev gu ij on uk ys .... sop
at ew ha is oy tu uh yn............ top
aq ey oh xu .... vop
av du ja op re to .... yop
ac ba do eg fe ib om pu ri .... zop
at ew ha is ki oy tu uh wo .... erp
ah ex hu ir ot pe ry ul wa.......... irp
ak ca em it ly ne ob uf yi zu .... asp
ac ba do eg fe ib ky om pu uw yx .... bup
ec iq ny ol su ub vo ze.............. cup
af hy ig ko oc uz vu yt zi .... dup
aj cy ef il oz ra uv .... gup
ak ca em it ly ne ob so yi zu......... hup
ax er iw je mi no ov ta up yu .... jup
az ep ju la ni ok ro ty us we yc...... kup
by en ge if ji os ru uj yd zo .... lup
at ew ha is ki oy tu uh wo yn .... mup
an eh go ip ka lu od py se um yl ..... nup
ap es ih ku of uc ym .... pup
ar ez id og po ux wy.............. rup
ek jo or te ug.................. ... sup
av du eq fy iz ja op ur .... tup
ab bi co da ev gu ij le my on uk ys... vup
am el fo nu sa .... yup
aw ho ic li wu xe .... zup
ax er iw je mi no ov sy ta up........ cyp
an eh fi go ip ka lu od py se um..... dyp
ag bu ce dy eb ik mo ox un xi za .... hyp
am el fo gi iv ke nu oj sa ud .... kyp
aj bo cy de ef il mu oz ra si uv ...... lyp
al by en ge if ji os ru uj va zo .... nyp
he in............................ ryp
as ek fu im jo or pa te ug wi......... typ

R

ag bu ce dy eb xi za .... bar
aj de ef il oz ra si uv .... car
ak ca em it ly ne uf yi zu .... dar
az ep ju la ni ok ro ty yc .... far
al en ge if ji ru uj yd zo .... gar
ix ow ut .... iar
am fo gi iv ke nu ud yb .... jar
an eh fi go ip ka lu od py se um yl... kar
ar cu ez id og ti ux wy .... lar
ek jo or pa te ug wi .... mar
av ci eq fy iz ja op re to ur .... nar
ab bi co da ev gu ij le my on uk ys... par
ax iw je mi no ov sy .... rar

Terminational Order.

R

ap es gy ih ku na of uc ym	sar
aw et ho ic vy xe ya	tar
ac ba do eg fe ib ky om pu ri uw yx	var
af be ed ga hy ig ko oc uz vu yt zi	war
at ew ha is ki oy tu uh wo	xar
ec ol su ub vo ze	yar
he in xo	zar
ac ba do eg fe ib ky ri uw yx	ber
ad ec iq ol ub vo ze	cer
af be ig oc uz vu yt zi	der
ag bu ce dy eb un xi za	fer
aj bo cy ef il oz uv	ger
ar cu id ma og po ti ux ve wy	her
ak em it ob so uf zu	jer
az ep ju ni ok ro ty us we yc	ker
by en ge if ji os uj va yd zo	ler
el fo gi iv ke nu oj sa ud yb	mer
an eh go ip ka od py se um yl	ner
ap ih me na of uc ym	per
he in oq xo	rer
aw et ho ic uq vy wu xe ya	ser
av du eq fy iz ja re ur	ter
bi da ev gu uk ys	ver
ex hu ir ot pe ry ul vi wa yo	wer
ay di ej ow ut	xer
as ek fu im or pa te ug wi	zer
ah ex hu ir ot pe ry vi wa yo	bir
fi ka py se um yl	cir
aw et ho ic li uq vy wu xe ya	dir
bi co da gu ij le on uk ys	fir
ax er iw je mi no ov sy ta up yu	gir
aj bo cy de ef il mu oz ra si uv	hir
af be ed ga hy ig ko oc uz vu yt zi	kir
at ha is ki oy tu uh wo yn	lir
av ci du fy iz ja op ur	mir
ay di ix ow qu ut xa ye	nir
ag bu dy eb ik mo ox un xi za	pir
az ju la ok ty we	sir
fu jo or te ug	tir
ar cu ez id ma og po ti ux wy	vir
am el fo gi iv ke nu oj sa ud yb	wir
ec fa hi ol su ub vo ze	zir
af be ed ga hy ig ko uz vu yt zi	bor
ag bu ce dy eb ik mo ox xi za	cor
aj bo cy ef il mu oz si uv	dor
ak ca ly ne ob uf yi zu	for
av ci du eq fy iz ja op re ur	gor
an fi go ip ka lu od py se um yl	hor
by en ge if os ru va yd	jor
am fo gi iv ke nu oj sa ud yb	kor
ay di ej ix ow qu ut xa	lor
cu ez id og po ux ve wy	mor
as ek fu im jo pa ug wi	nor
az ep ju la ni ok ro ty we	por
ab bi da ev gu ij le my uk	ror
ax er iw je no ov sy ta up	sor
ba eg ky om ri uw yx	tor
at is oy yn	uor

R

ap es gy ih ku lo me of pi uc ym	vor
ad fa hi iq ny ol su ub vo ze	wor
aw ho ic li uq wu ya	xor
ah ex hu ir ot pe wa	zor
ec iq ny ol ub vo ze	bur
af be ed ga hy ig ko uz yt zi	cur
az ep ju la ni ok ty we yc	dur
aj bo cy de ef il mu oz ra si uv	fur
ak ca em it ly ne ob uf yi zu	gur
ax iw je mi ov sy up yu	hur
at ew ha is ki tu uh	jur
el fo gi iv ke nu oj sa ud yb	lur
an eh fi go ip ka lu od py um	mur
ar cu ez id ma og po ti ux ve wy	nur
al en yd	our
as ek fu jo pa te ug wi	pur
bi co da gu ij le my on uk	sur
ag bu ce dy eb ik mo un xi za	tur
ah ex hu ir ot pe ry ul vi wa	vur
ci du fy ja to ur	xur
ac ba do eg fe ib pu ri	zur
aw et ho ic li xe	cyr
he in oq	fyr
af be ga ko oc uz vu	gyr
ad ec fa ny ol su ub vo	hyr
aj bo cy de ef il mu oz ra si uv	kyr
ak ca em it ly ne ob so uf zu	lyr
az ep ju la ni ok ro ty us we	myr
al by en ge if ru va zo	pyr
ac ba do eg fe ib om pu ri uw	syr
ap es gy ih ku lo me na of pi uc	tyr

S

af be ed ga hy ig ko uz vu zi	bas
ag bu ce dy eb xi za	cas
aj bo cy de ef il mu si uv	das
ak ca em it ly ob uf yi zu	fas
av ci du eq fy iz ja op re ur	gas
in oq xo	has
al by en ge if os ru va yd	jas
am el fo gi iv ke nu oj sa ud yb	kas
an eh fi go ip ka lu od py um	las
ar cu ez id og po ti ux ve wy	mas
as fu im or pa te ug wi	nas
az ep ju la ni ok ro ty we	pas
ab bi da ev gu ij le my uk	ras
ax iw je no ov sy ta up yu	sas
ad ec fa hi iq ny ol su ub vo ze	tas
ap es gy ih ku lo me of pi uc ym	vas
aq ey oh xu	was
ay di ej ow ut	xas
at ew ha is	yas
ex hu pe wa	zas
at ew ha is ki oy tu uh yn	abs
ag bu ce dy eb ik mo un xi za	ebs
af be ga hy ig ko oc uz vu yt	ibs
ac ba do eg fe ib ky om pu ri uw yx	obs
am el fo gi iv oj sa ud yb	ubs

S

ab bi co da ev gu ij le my on uk ys	acs
am el fo iv ke nu sa ud	ecs
aj bo cy de ef il mu ra uv	ics
at ew ha is ki oy tu uh wo yn	ocs
af ed ga hy ig ko oc uz vu yt zi	ads
av du eq fy iz ja op re to ur	ids
aj bo cy de ef il mu oz ra si uv	bes
ak ca em it ly ne ob so uf zu	ces
ci eq fy iz op re	des
ji ru va zo	fes
am el fo iv nu oj sa ud yb	ges
an eh fi go ip ka lu od se um	jes
ar cu ez id ma og po ti ux ve wy	kes
as ek fu jo or te wi	les
ap es gy ih ku lo me of pi uc ym	mes
ab ev gu le my on uk ys	nes
ax iw je mi no ov sy up yu	pes
ep ju ni ok ro us we yc	res
ac ba eg fe ib ky om pu uw yx	ses
ay di ej ix ow qu ut xa ye	tes
ad ec fa iq ny ol su ub vo ze	ves
af be ed ga hy ig ko oc uz vu yt zi	wes
in xo	yes
aj bo cy de ef il mu oz ra si uv	ahs
av ci eq fy iz ja op	bis
ax je sy ta up	cis
az ep ju la ni ok ro ty we yc	dis
ah ex hu ir ot pe ry ul vi wa yo	fis
af be ga hy ig ko oc uz vu yt	gis
cu id ma og po ti ux wy	his
ak ca em it ne ob so uf zu	jis
aj bo cy de ef il mu oz ra uv	kis
ap es gy ih lo na of uc ym	lis
aw et ho ic li uq vy wu xe ya	mis
ag bu ce dy za	nis
aq	ois
ay ej ix ow qu ut xa	pis
ek fu im jo te ug wi	sis
am el fo gi ke nu oj ud yb	tis
at ew is ki oy tu uh yn	vis
al en ge if ji ru uj yd zo	wis
an fi go ip ka lu od um	xis
he in xo	zis
he in oq xo	oks
az ep ju la ni ok ro ty us yc	als
oq xo	els
ax er iw je ov sy up yu	ils
ab bi da ev gu ij le my on uk ys	ols
ar cu id og po ve	yls
av ci du eq fy iz ja op to ur	ams
as ek fu im jo or pa ug wi	ems
ah ex hu ir ot pe ry ul wa yo	ims
ap es gy ih ku me na of pi uc ym	oms
ay di ej ix ut xa ye	ums
el fo gi iv ke nu oj sa ud yb	ans
ac ba do eg fe ib ky pu ri uw yx	ens
ay ej ix ow ut xa ye	ins
av ci du eq fy iz ja op to ur	ons

Terminational Order.

S

el iv ke oj sa ud yb	bos
ay di ej ix ow qu xa	cos
er iw je ov sy ta yu	dos
aw et ic li uq vy wu xe ya	fos
as ek im or te ug	gos
aq xu	hos
ap es ih ku lo me na of pi uc	jos
co da ev gu ij le uk ys	kos
av ci du eq fy iz ja op re ur	los
az ep ju la ok ro ty we yc	mos
ah hu ir ot ry ul yo	nos
ad ec iq ny ol ub vo	pos
be ed hy ig ko oc uz vu yt zi	ros
ag bu ce dy eb ik ox un xi za	sos
ez id og ux ve	tos
aj bo cy de ef il mu oz ra si uv	vos
al ge ru va zo	zos
ah ex hu ir ot ry ul vi wa yo	eps
az ep ju la ok ro ty we yc	ips
an eh fi go ip ka lu od py se um yl	ops
as ek im jo or pa te ug wi	ups
er iw mi no ov sy ta yu	ars
aw et ho ic li uq vy wu xe ya	ers
in oq xo	ats
av ci eq fy iz ja op re ur	ets
ac eg fe ib ky uw yx	its
ak ca em it ly ne ob so uf yi zu	ots
ex ir ot	aus
ax er je mi ov up yu	bus
ac eg fe ib ky pu ri uw yx	cus
ec iq ny su vo	dus
af be ed ga hy ig ko oc uz vu yt zi	fus
ce dy eb ik mo ox un xi	gus
ix ow qu ut xa ye	hus
as ek fu jo or pa ug wi	jus
ak ca em it ly ne ob so uf yi zu	kus
et ic li uq vy wu xe ya	lus
en if uj yd zo	mus
am el fo gi iv ke nu oj ud yb	nus
an eh go ip ka py se um	pus
av ci fy re ur	rus
cu ez id ma og po ux ve	sus
gy ku uc ym	tus
az ep ju la ni ok ro ty us we yc	vus
ab ev le on uk	yus
at ew ha ki uh wo	zus
al by en if os uj yd	ews
ah ex ir ot pe ry ul vi wa	ows
az ep ju la ni ok ro ty us	bys
al by en ge if ji os uj va zo	cys
aw et ho ic li wu xe	dys
am el fo gi iv ke nu sa ud	fys
aj bo cy il mu ra si uv	hys
ax er iw je mi no ov sy ta up	kys
ej ix ow ut xa	lys
ek fu jo or pa te ug wi	mys
ap es ih ku lo me na of pi uc	nys
ag bu ce dy eb ik mo ox un za	oys
ah hu ot ul	rys
ak ca em it ly ne ob so uf zu	sys
ci du fy iz ja re to ur	tys

T

ah ex ir ot pe ry ul vi wa yo	bat
aw et ho ic li uq vy wu ya	cat
ay di ej ix ow qu ut xa	dat
ak em ob uf	eat
ap es gy ih ku lo na of pi uc ym	fat
ab bi co da ev gu ij my on uk ys	gat
al en ge if ji ru uj yd zo	hat
bo de ef il mu ra si uv	jat
er iw je mi no ov ta yu	lat
at ew ha is ki oy tu uh yn	mat
aq oh xu	nat
in oq xo	pat
ec iq ny ol ub vo ze	rat
av du eq fy iz ja op re ur	sat
af be ed ga hy ig ko oc uz vu yt	tat
an eh fi go ip ka lu od py se um yl	vat
ac ba do eg ib ky om pu ri yx	wat
ag ce eb ik mo ox un za	yat
ar cu ez ma og po ti ux	zat
ec iq ny ol su ub vo ze	act
as ek fu im jo or pa te ug wi	ect
ak ca em it ly ne ob uf yi zu	bet
ar cu ez id ma og po ti ux ve wy	cet
at ew ha is ki oy tu uh wo yu	det
ay di ej ix ow qu ut xa ye	fet
ah ex ir ot ry ul vi yo	get
an go ka lu od py se um yl	het
al en ge if ru yd zo	jet
aw et ho ic li uq vy wu xe ya	ket
ap es gy ih ku lo na of pi uc ym	let
in oq xo	met
ek fu jo pa ug wi	net
av ci du eq fy iz op re	pet
aq oh xu	ret
az ep ju la ok ro ty we yc	tet
ax er iw je mi no ov sy ta up yu	vet
bi co da ev gu ij le on uk ys	wet
el fo iv ke nu sa ud	yet
ac ba do eg fe ib pu ri	zet
ak ca em it ly ne ob uf yi zu	aft
at ew ha is ki oy tu uh yn	eft
ag bu ce dy eb un xi za	oft
ah ex ry ul	ait
as ek fu jo pa te ug wi	bit
af be ed ga hy ig ko oc uz vu	cit
in oq xo	dit
al en ge if ji ru uj yd zo	fit
av ci du eq fy iz ja op re ur	git
ba do eg ib ky om pu uw yx	hit
ag bu ce dy eb un xi za	kit
am el fo gi iv ke nu oj sa ud yb	lit
ar cu ez id ma og po ti ux wy	mit
az ep ju la ni ok ro ty we yc	nit
fi go ip ka lu od py se um yl	pit
ax er iw je mi ov sy up yu	rit
ak ca em it ly ne ob uf zu	sit
bi co ev gu ij le on uk ys	tit
aj bo cy de ef il mu oz ra si uv	vit
iq ny ol su ub ze	wit
ap ku lo of pi	zit
aq ey oh xu	alt
av ci du eq fy iz ja op re to	olt
ab bi co da ev ij le my on uk ys	ult
aj bo cy de ef il mu oz ra si uv	amt
ac ba do eg ib ky om pu ri uw yx	ant
bi co da gu ij on uk ys	ent
el fo gi iv ke nu oj sa ud yb	ont
ce dy eb ik ox un xi za	unt
ay di ej ix ow qu ut xa	bot
ah ex ir ot pe ry ul wa yo	cot
aw et ho li uq vy wu xe ya	dot
an fi go ip ka lu od py se um yl	fot
cu id og po ti ux ve wy	got
ec iq ny ol su ub vo ze	hot
ax er iw je mi no ov ta up yu	jot
at ew is ki oy tu uh yn	kot
en ge ji ru yd zo	lot
av ci du eq fy iz ja op re to ur	mot
co da ev gu ij le on uk	not
ac ba do eg ib ky pu uw yx	pot
ih ku lo na of pi uc ym	rot
in oq xo	sot
aj bo de ef il mu oz	tot
ag bu ce dy eb ik mo ox un xi za	vot
af be ed ga hy ko oc uz vu yt zi	wot
ak ca em ne uf zu	yot
ep ju la ni ok ro we	zot
er iw je mi no ov sy ta yu	apt
aj bo cy de ef il mu oz ra si uv	ept
az ju la ni ok ro ty we yc	art
bu ce dy eb ik mo ox un xi za	ert
ad ec fa hi iq ny ol su ub ze	ort
af be ga hy ig ko oc uz vu yt	est
bu ce dy eb ik ox un za	ist
aj bo cy de ef il mu oz ra si uv	ost
ec hi iq ny ol ub vo ze	ust
an eh fi go ip ka lu od py se um yl	cut
al by en if ji os ru uj va yd zo	dut
ar cu ez id ma og po ti ux ve wy	fut
as ek fu jo or pa te ug wi	gut
bi co da ev gu le uk ys	hut
ap es ih ku lo na of pi uc	jut
ah ex hu ir ot pe ry ul vi wa yo	kut
ak ca em it ly ne ob uf yi zu	lut
am fo gi iv ke nu oj sa ud yb	mut
at ew ha is ki oy tu uh yn	nut
ay ej ix ut	out
ag bu dy eb xi za	put
af be ed ga hy ig ko oc uz vu zi	rut
aq oh xu	sut
ec iq ny ol su ub vo ze	tut

Terminational Order.

T

ba eg ib om pu	yut
ay di ej ix ow qu ut xa	axt
ax er iw mi no ov sy ta up	byt
aq ey oh xu	cyt
am el fo gi iv ke nu oj sa ud	dyt
ad ec fa hi iq ny ol su ub vo ze	lyt
ap es gy ih ku lo me na of pi uc	myt
ah ex hu ir ot pe ry vi wa	nyt
ab bi co da ev le on uk	pyt
an eh fi go ip ka od py se um	tyt
az ep ju la ni ok ro us	vyt

U

af be ed ga hy ig ko oc uz vu yt zi	abu
ac ba do eg fe ib ky om pu ri uw yx	ebu
ap es gy ih ku lo me na of uc ym	ibu
ab bi co da ev gu ij le my on uk ys	obu
at ew ha is ki oy uh wo yn	ubu
av ci du eq fy iz ja op re to ur	acu
al en ge if ji ru yd zo	ecu
ab co da ev gu le my on uk	icu
ep ju la ni ok ro ty us we yc	ocu
ag bu dy eb ik mo ox un xi za	scu
ar ez id ma og po ti ve wy	ucu
an eh fi go ip ka lu od py se um yl	adu
as ek fu jo or pa te	idu
ag bu ce dy eb ik mo ox un xi za	odu
ay di ej ix ow ut xa ye	udu
al by en ge if ji os ru uj va yd zo	afu
ej ix ow ut xa ye	efu
ar cu ez id ma og po ux ve wy	ifu
ad ec fa hi iq ny ol su ub vo ze	ofu
av ci eq fy iz ja op re to ur	ufu
as ek fu im jo or pa te ug	egu
ac ba do eg fe ib ky om pu uw yx	igu
ap es gy ih lo me na of pi uc ym	ugu
ak ca em it ly ne ob uf yi zu	ahu
ex hu ir ot pe ry ul wa yo	shu
ac ba do eg fe ib ky om pu ri uw yx	aju
el fo gi iv ke nu sa ud	eju
ax er iw je mi no ov sy ta up yu	aku
ar cu ez id ma po ti ux ve wy	eku
an eh go ip ka lu od py se um yl	iku
ac ba do eg fe ib ky om pu ri uw yx	oku
ak ca em it ob so uf yi	uku
at ew ha is ki oy tu uh wo yn	alu
ab bi co da ev gu ij le my on uk ys	blu
en ji ru uj va yd zo	clu
av ci du eq fy iz ja ur	elu
ad ec fa hi iq ny ol su ub vo ze	flu
aw et ho ic li vy wu xe	glu
ay ej ix ut xa	ilu
aj bo cy de ef il mu oz ra si uv	klu
as ek fu im jo or pa te ug wi	olu
am el fo gi iv ke nu oj sa ud yb	plu
an eh fi ip ka od py se um	ulu
az ep ju la ni ok ro ty us we yc	amu
ak ca em it ly ne ob uf yi zu	emu
ag bu ce dy eb ik mo ox un za	imu
ar cu ez id ma og po ti ux ve wy	omu
er iw je mi no ov sy ta	umu
ay di ej ix ow ut xa ye	anu
ax er iw je mi no ov sy ta up yu	enu
av du eq fy iz ja op re to ur	inu
aj bo cy de ef il mu oz ra si uv	onu
az la ni ok ro ty us we	unu
an eh fi go ip ka lu od py se um yl	epu
ak ca em it ly ne ob so uf zu	ipu
am el gi iv ke nu oj sa ud yb	opu
ay di ej ix ow qu ut xa ye	spu
ah ex ir ry ul vi wa	upu
in oq	cru
ay di ej ix ow qu ut xa ye	dru
ap es gy ih ku me na of pi ym	eru
ac ba do eg fe ib ky om pu ri uw yx	fru
ar cu ez id ma og ti ux ve wy	gru
al by en ge if os ru uj va yd zo	iru
av ci du eq fy iz ja op re to ur	kru
aw et ho ic li uq vy wu xe ya	pru
af be ed ga hy ig ko oc vu zi	tru
ab bi co da ev gu ij le my on uk	esu
ap es gy ih ku lo me na of uc ym	atu
ex hu ir ot pe ry ul vi wa yo	etu
ax er iw je no ov sy ta up yu	itu
eh fi go ip ka lu od py se um yl	otu
at ew ha is oy tu wo yn	stu
ba do eg fe ib ky om ri uw yx	utu
am el fo gi iv ke nu oj sa ud yb	avu
aj bo cy de ef il mu oz ra si uv	evu
ax er iw je mi no ov sy ta up yu	ovu
aw et ho ic li uq xe ya	nvu
ar cu ez id ma og po ti ve wy	axu
az ep ju ni ok ro ty we yc	exu
aw et ho wu ya	ixu
ap es gy ih ku lo me na of pi uc ym	oxu
as ek im jo or pa ug	uxu
ag bu ce dy eb ik mo ox un za	ezu
aj bo cy de ef il mu oz ra uv	izu
aw et ho ic li vy wu xe ya	ozu
al by en ge if ji os va zo	uzu

V

ad ec fa hi iq ny ol su ub vo ze	bav
af be ed ga hy ig ko oc uz zi	cav
ag bu ce dy eb ik mo ox un xi za	dav
aj bo cy de ef il mu oz ra si uv	fav
ak ca em it ly ne ob so uf yi zu	gav
ac ba do eg fe ib ky om pu ri uw	hav
ah ex hu ir pe ul vi wa yo	jav
al by en ge if ji os ru uj va yd zo	kav
at ew ha is ki oy tu uh wo yn	lav
ar cu ez id ma og po ti ux ve wy	nav
as ek fu im jo or pa te ug wi	pav
es gy ih ku lo me na of pi uc ym	rav
ab bi co da gu ij le my on uk ys	sav
am el fo gi iv ke nu oj sa ud yb	tav
ax er iw je mi no ov sy ta up yu	vav
az ep ju la ni ok ro ty us we yc	wav
he in oq	xav
av ci du eq fy iz ja op re to ur	bev
ac ba do eg fe ib ky om pu ri uw yx	dev
ad ec fa hi iq ny ol su ub vo	fev
be ed ga hy ig ko oc uz vu yt zi	gev
aj bo cy de ef il mu oz ra si uv	kev
ay di ej ix ow qu ut xa ye	lev
ax er iw je mi no ov sy ta up yu	mev
al by en ge if ji os ru va yd zo	nev
ar cu ez id ma og ti ux ve wy	pev
an eh fi go ip ka lu od py se um yl	rev
as ek fu im jo or pa te ug wi	sev
al by en ge if os ru uj va yd zo	biv
ak ca em it ly ne ob so uf yi.	civ
am fo gi iv ke nu oj sa yb	div
as ek fu im jo or pa te ug wi	giv
av ci du eq fy iz ja op to ur	hiv
ag bu ce dy eb ik mo ox un xi za	liv
ap es gy ih ku lo me na of pi uc ym	niv
ac ba do eg fe ib ky om pu ri uw yx	piv
ju la ni ok ro ty us we yc	riv
aq xu	siv
aj bo cy de ef il mu oz ra si uv	tiv
ad ec fa hi iq ny ol su ub vo ze	viv
an eh fi go ip ka lu od py se um	wiv
ak ca em it ly ne ob so uf yi zu	elv
ad ec hi iq ny ol su ub vo ze	cov
af be ed ga hy ig ko oc uz vu yt zi	dov
aj bo cy de ef il mu oz ra si uv	gov
al by en ge os ru uj va yd zo	hov
ax er iw je mi no ov sy ta up yu	lov
am el fo gi iv ke nu oj sa ud yb	mov
an fi go ip ka lu od py se um yl	nov
ap es ih ku me na of pi uc ym	pov
ar cu ez id ma og po ti ux ve wy	rov
ay di ix ow qu ut xa ye	sov
ab bi co da ev gu ij le my on uk ys	vov
et ho ic li uq vy wu xe ya	wov
as ek fu im jo or pa te ug wi	duv
he in oq xo	kuv
ac ba do eg fe ib ky om pu ri uw yx	luv
ak ca em ly ne ob uf zu	suv

W

aw ho ic uq vy wu xe ya	baw
an eh fi go ip ka lu od py se um yl	caw
az ep ju la ni ok ro ty we yc	daw
ar cu ez id ma po ti ux ve wy	faw
ah ex hu ir ot pe ry ul vi wa yo	gaw
bi co da ev gu le uk ys	haw
av du ja op re ur	jaw
at ew ha is ki oy tu uh yn	kaw
am el fo gi iv ke nu oj sa ud yb	law
ay di ej ix ow qu ut xa	maw
he in oq xo	naw
ag bu dy eb un xi za	paw

[redacted]

Terminational Order.

W

aj bo cy de ef il mu oz ra si uv ...... raw
ak ca em ly ne ob uf yi zu .... saw
ac ba do eg ib ky om pu ri uw yx .... taw
al by en ge ji os ru uj va yd zo ...... waw
af ed ga ko vu .... yaw
aw et ho ic li uq vy wu xe ya........ dew
am el fo gi iv ke nu oj sa ud yb...... few
eh fi go ip ka lu od py se um yl...... gew
af be ed ga hy ig ko uz vu zi .... hew
bo de ef il mu ra si uv .... jew
by en ge if ji os ru uj va yd zo....... kew
he in oq........................... lew
ah ex hu ir ot pe ry ul vi wa yo...... mew
ak ca em ly ne ob uf yi zu .... new
ay di ej ix ow qu ut xa .... pew
ac ba do eg fe ib ky om pu ri uw yx .. rew
ec iq ny ol su ub vo ze .... sew
ag bu ce dy eb un xi za .... tew
co da ev gu le on uk .... yew
ah ex ir ot pe ry ul vi wa yo......... bow
er iw je no ov ta .... cow
aq ey oh xu .... dow
ay di ej ix ow qu ut xa ye .... gow
ap es gy ku lo na of pi uc ym........ how
at ew ha is ki tu uh wo .... jow
aw et ho ic li uq vy wu xe ya .... kow
an eh fi ip ka lu od py se um .... low
in oq xo........................... mow
aj bo cy de ef il mu oz ra si uv....... pow
iq ny ol su ub vo ze .... row
ar cu ma og po ti ux ve wy.......... sow
bi co da ev gu ij le on uk ys......... tow
az ep ju la ni ok ro ty we yc......... vow
av ci du fy iz ja op re ur............ wow
ac ba eg fe ib pu.................. yow
he oq xo.......................... huw
af be ed ga ko oc vu yt zi .... juw
av ci du eq fy iz ja op re ur .... luw
ay di ej ix ow ut xa ye............. ruw

X

ac ba do eg fe ib ky om pu uw yx .... bax
ad ec iq ny ol ub vo ze.............. cax
af hy ig ko oc uz vu yt zi .... dax
ag bu ce dy eb un xi za .... fax
aj cy ef il oz ra uv .... gax
az ep ju la ni ok ro ty us we yc...... hax
ew .... iax
en ge if ji ru uj yd zo .... lax
am fo gi iv ke nu oj sa ud yb .... max
an eh go ip ka lu od py se um yl..... nax
ap es ih ku of uc ym .... pax
ar ez id og po wy .... rax
ix ow qu ut xa..................... sax
av du eq fy iz ja op ur .... tax
er iw mi ov sy yu .... wax
as ek fu jo or pa ug wi.............. zax
at ew ha is oy tu uh wo yn.......... bex

X

az ep la ok ro ty us we yc .... dex
av ci du eq fy iz ja op re ur .... fex
ag bu ce dy eb un xi za............. hex
ah ex hu ir ot pe ry ul vi wa yo ..... kex
ak em it ly ne ob uf yi zu .... lex
ab bi co da ev gu ij le my on uk ys... mex
aw et ho ic li uq vy wu xe ya .... nex
as ek fu im jo or pa te ug wi .... pex
es gy ih ku lo na of pi uc ym........ rex
ac ba do eg fe ib ky om pu ri uw yx.. tex
ay di ej ix ow qu ut xa .... vex
aj bo cy de ef il mu oz ra si uv ...... wex
ec ol su ub vo ze.................. yex
as ek or ug .... aix
am gi iv ke nu oj sa ud yb .... bix
ec iq ny ol su ub vo ze .... dix
ak ca em ly ne ob uf yi zu .... fix
ab co da ev gu ij le my on ys lix
aj bo cy ef il mu oz ra uv .... mix
ac ba do ib ky om pu ri uw yx....... nix
az ep ju la ni ok ro ty we. .... pix
aw et ho ic vy xe ya............... rix
ap es gy ih me pi uc .... tix
an eh fi go ip ka lu od py se um yl... vix
ar ez id og po ti ux ve wy .... box
as ek fu jo pa te ug wi............... cox
ap es gy ih ku lo me na of pi uc ym... dox
bi co da ev gu ij le on uk ys......... fox
aj bo cy de ef il mu oz ra si uv....... hox
ad ec fa iq ny ol su ub vo ze .... lox
af be ed ga hy ig ko oc uz vu zi...... mox
ag ce dy eb ik mo ox un za .... nox
em it ly ne ob so uf yi zu........... rox
al by en ge if ru uj yd zo............ sox
eh go ip ka lu od py se um yl........ tox
am el fo gi iv ke nu oj sa ud yb ...... vox
ay di ej ix ow qu xa .... wox
an eh fi go ip ka lu od py se um yl ... arx
al by en if ji os ru uj va zo .... erx
ad iq ny ol ub .... aux
ag bu ce dy eb un xi za,............. bux
aj bo de ef il oz ra si uv .... cux
ak ca em it ly ne ob so uf yi zu ...... dux
aw et ic .... eux
az ep ju la ni ok ro ty we yc......... fux
am el fo gi iv ke nu oj sa ud yb ...... hux
an eh fi go ip ka lu od se um .... jux
ay di ej ix ow ut xa................ kux
ar cu ez id ma og po ti ve wy .... lux
as ek fu im jo or pa te ug wi .... mux
av ci du eq fy iz ja op re ur .... nux
ah ir ot ry ul..................... oux
ap es gy ih lo me na of pi uc .... sux
at ew ha is ki oy tu uh wo yn........ tux
er je no ov ta .... yux
an eh fi go ip ka lu od se um .... byx
ak ca em it ly ne ob so uf zu .... cyx
ab bi co da ev gu ij le my on uk...... dyx

X

aw et ho ic li wu xe .... kyx
az ep ju la ni ok ro ty us we .... lyx
di ej ix ow ut xa .... myx
as ek fu im jo or pa te ug wi......... nyx
el fo iv ke nu sa ud................ pyx
ax er iw je mi no ov sy ta up .... ryx
ar cu ez id ma og po ti ux ve .... tyx

Y

in oq xo.......................... bay
ar cu ez id ma og po ux ve wy .... cay
as ek fu jo or pa te ug wi .... day
av ci du eq fy iz ja op re ur .... fay
ag bu ce dy eb un xi za............. gay
an fi go ip ka lu od py se............ hay
er iw mi no ov ta .... jay
ah ex hu ir ot pe ry ul vi wa......... kay
ac eg ib ky om pu ri uw............. lay
aw et ho ic li uq wu xe .... may
ec ny su ub vo ze.................. nay
at ew ha is ki oy tu................ ray
aq ey oh xu .... tay
al by en ge if ji os ru uj va yd zo..... vay
aj bo cy de ef il mu oz ra si uv....... way
cu ez ma og po ti ux ve .... aby
et wu xe ya....................... eby
ax er iw je no ov sy ta up .... iby
at ew ha is ki oy tu uh wo........... oby
av iz ja op to ur .... uby
an eh fi go ip ka lu py se um .... acy
iw no ov ta up yu .... ecy
ny ol su ub vo ze.................. icy
im jo or pa te ug wi .... ucy
ad ec fa hi ol ny su ub vo ze .... ady
ak ca em it ly ob so uf zu .... edy
az ep ju la ok ro ty us we .... idy
al by en ge if ji os ru uj va zo........ ody
aj bo cy de ef il oz ra si uv .... udy
ay di ej ix ow ut xa .... bey
az ep ju la ni ok ro ty us............ cey
al by en if ji os ru uj va zo .... dey
aj bo cy de ef il mu oz ra si uv....... fey
am el fo gi iv ke nu oj sa ud .... hey
as ek fu jo or pa te ug wi............ key
ah ex hu ir ot pe ul wa .... ley
at ew ha is ki oy tu wo .... mey
ax er iw je mi no ov sy ta up .... ney
aw et ho ic li wu xe .... rey
he in oq xo .... sey
ad ec fa hi iq ny su ub vo ze.......... tey
av ci du eq fy iz ja op re to ur .... vey
ap es ih ku lo na of pi uc .... wey
af be ed ga ko oc vu................ zey
ep ju la ni ok ro ty us we .... afy
cu id ma og po ti ux ve wy .... efy
ap es gy ih ku lo me na of uc........ ify
ak ca em it ly ne ob so yi zu......... ofy
aw et ho ic li xe................... ufy

Terminational Order.

Y

ap es ku lo me na of pi uc	agy
he in oq xo	egy
an eh go ip ka lu od se um	igy
ab bi co da ev gu ij le my on uk	ogy
as ek fu jo or pa te ug	ahy
aj bo de ef il mu oz ra si uv	chy
av ci du fy iz ja op re to ur	phy
am el fo gi iv ke nu oj sa ud	shy
bi co da ev gu ij le on uk	thy
et ic li wu xe	why
ap es gy ih ku lo me na of pi uc	eky
ej ix ow ut xa	iky
af be ed ga hy ig ko oc uz vu zi	oky
aw et ho ic li wu xe	sky
al by en ge if ji os uj va zo	uky
ac ba do eg fe ib ky om pu ri uw	aly
ag bu ce dy eb ik mo ox un xi za	bly
in oq xo	cly
az ep ju la ni ok ro we	dly
ix ow qu ut xa	ely
af be ed ga hy ig ko oc uz vu yt zi	fly
ap es gy ih ku lo me na of pi uc	gly
as ek fu im jo or pa te ug	ily
ad ec fa hi ol su ub vo ze	oly
et ho ic li wu xe	ply
bi co da ev gu ij le on	sly
am el fo gi iv oj sa ud yb	uly
ax er iw je mi no ov ta up	zly
ak ca em it ly ne ob so uf zu	amy
aj bo cy ef mu oz ra si uv	emy
ew ha is oy tu uh wo yn	imy
ax er iw je mi no ov sy ta up	omy
ab bi co da ev ij my on uk ys	umy
er iw je mi no ov sy ta	any
av ci du eq fy iz ja op re to ur	eny
ak ca em it ly ne ob so uf zu	iny
di ej ix ow qu ut xa	ony
aq ey oh xu	sny
al en ge if ji ru uj zo	boy
an eh fi go ip ka lu od py se um	coy
ar cu id ma og po ti ux ve wy	foy
ac ba do eg ib ky om pu ri uw	hoy
ec ol su ub vo ze	joy
he in oq xo	koy
ah ex ir ot pe ry ul vi wa	moy
aw et ho ic li wu xe	noy
ay di ej ix ow qu ut xa	poy
aq oh xu	roy
av ci du eq fy iz ja op re ur	soy
af be ed ga hy ig ko oc uz zi	toy
ax er iw je mi no ov sy ta up	voy
ak ca em it ne ob so zu	zoy
ab bi co da ev gu ij le my on uk	apy
as ek fu im jo or pa te ug wi	epy
av du op re ur	ipy
ar cu ez id ma og po ti ux ve	opy
an fi go ip ka lu od py se um	spy
ah ex hu ir pe ry ul wa	ary

Y

as ek fu im jo or pa te ug wi	bry
ac ba do eg ib ky om pu ri uw	dry
an go ip ka lu od py se um	ery
er iw je mi no ov sy ta	fry
az ep ju la ni ok ro ty us we	gry
al by en ge if ji os ru uj va zo	kry
es ku lo me na of uc	ory
af be ed ga hy ig ko oc uz vu zi	pry
ak ca em it ly ne ob uf zu	try
aq ey oh	ury
av ci du fy iz ja op re ur	wry
ay di ej ix ow qu ut xa	asy
ad ec fa hi iq ny ol su ub vo ze	esy
ab co ev gu ij le uk	isy
az ep ju la ni ok ro us we	osy
af be ed ga hy ig ko oc uz vu zi	aty
aj bo cy ef il mu ra	ity
aq ey oh xu	oty
ar cu id og po ti ux ve	sty
ax er iw je mi no ov sy ta up	duy
ah ex ir ot pe ry ul vi wa	guy
as ek fu im jo or pa te ug wi	huy
in oq	muy
ew ha is ki oy tu uh wo	puy
ac ba do eg fe ib om pu ri uw	tuy
ar cu ez id ma og po	ivy
av ci du eq fy iz ja op re to ur	axy
aj bo cy de ef il mu oz ra si uv	oxy

Z

ax er iw je mi no ov sy up yu	baz
ac ba do eg fe ib ky om pu ri uw yx	caz
ad ec fa hi iq ny ol su ub vo ze	daz
af be ed ga hy ig ko oc uz vu yt zi	faz
ew ha ki oy tu uh yn	gaz
ag bu ce dy eb ik mo ox un xi za	haz
di ix ow ut xa ye	jaz
ak ca em it ly ne ob so uf yi zu	kaz
aw et ho ic li uq vy wı xe ya	laz
al by en ge if ji os ru va yd zo	maz
am el fo gi iv ke nu oj sa ud yb	naz
an eh fi go ip ka lu od py se um yl	paz
ek fu im jo or pa te ug wi	raz
cu ez id ma og po ti ux ve wy	saz
ap es gy ih ku me na of pi uc	taz
av ci du eq fy iz ja op re to ur	vaz
ab bi co da ev gu ij le my on uk ys	waz
bo	yaz
ar cu ez id ma og po ti ux ve wy	adz
ag bu ce dy eb ik mo ox un xi za	bez
ak ca em it ly ne ob so uf yi zu	dez
az ep ju la ni ok ro ty we yc	fez
by en ge if ji os ru uj va yd zo	gez
at ew ha is ki oy tu yn	hez
am el fo gi iv ke nu sa ud yb	jez
cu ez id ma og po ti ux wy	lez
as ek fu jo or te ug wi	mez
av ci du eq fy iz ja re to ur	nez
ab bi co da ev gu ij le my on uk ys	pez

Z

ad fa hi ny ol su ub vo ze	rez
ap es gy ku lo me na of pi uc ym	sez
af be ed ga hy ig ko oc uz vu yt zi	tez
ac do eg fe ib ky om pu ri uw yx	vez
ix ow xa	yez
ex hu ot vi wa	zez
ak ca em it ly ne ob so uf yi zu	biz
av ci du eq fy ja op re to ur	ciz
af be ed ga hy ig ko oc uz vu yt	diz
aj bo cy de ef il mu oz ra si uv	fiz
am el fo gi iv ke nu oj sa yb	giz
ay di ix ut xa	hiz
by en ge if ji os ru uj va yd zo	kiz
ad ec fa hi iq ny ol su ub vo ze	liz
ap es gy ih ku lo me na of uc ym	miz
at ew ha is ki oy tu wo yn	niz
ax iw je mi no ov sy up yu	piz
an fi go ip ka lu od py se um yl	riz
ba do eg fe ib ky om pu ri uw yx	siz
ag bu ce dy eb ik mo ox un xi za	tiz
ex hu ot ry ul vi wa yo	wiz
aw et ho ic li wu	ziz
aj bo cy de ef il mu oz ra si uv	elz
ag bu ce dy eb ik mo ox un xi za	anz
ad fa iq ny ol su ub vo ze	inz
ap es gy ih ku lo me na of pi uc ym	onz
ab bi co da ev ij le my on uk ys	unz
az ep ju la ni ok ro ty we yc	boz
ar cu ez id ma og po ux ve wy	coz
as ek fu im jo or pa te ug wi	doz
ap es gy ih ku lo na of pi ym	foz
ab bi co da ev gu ij le my on uk	goz
am fo iv ke nu oj sa yb	hoz
av ci du eq fy iz ja op re to ur	koz
ac ba do eg fe ib ky om pu ri uw yx	loz
ad ec fa hi iq ny su ub vo	moz
af be ed ga hy ig ko oc uz vu yt zi	noz
ag bu dy eb ik mo ox un xi za	poz
aj bo cy de ef il mu oz si	roz
ak ca em ly ne ob uf zu	soz
al by en ge if ji os ru uj va yd zo	voz
an go ip ka lu um	zoz
ah ex ir ot pe ry ul wa yo	erz
aq ey xu	etz
ah ex hu ot pe ry ul wa yo	itz
at ew ha is ki oy tu uh wo	otz
as ek fu im jo or pa te ug wi	buz
ap es gy ih ku lo me na of pi uc ym	cuz
ab co da ev gu ij le my on uk ys	duz
av ci du eq fy iz ja op re to ur	fuz
az ep ju la ni ok ro ty we yc	guz
ar cu ez id ma og po ti ve wy	huz
af be ed ga hy ig ko oc uz vu yt zi	luz
ag bu ce dy eb ik mo ox un xi za	muz
aj bo cy de ef il mu oz ra si uv	nuz
ak em it ly ob so uf zu	puz
al by en ge if os ru uj va yd zo	ruz
am el fo gi iv ke nu oj sa ud yb	suz
ah ex ir ot pe ry ul vi wa yo	tuz
ax er je mi no ov ta up	zuz

[redacted]

DETECTORS.

Detector.

A

aa ek ent rt **fl ft**
ab ems pee pi ws wei wie **w3**
ac egn emr pn pte wae wen wr **eb f2 xe**
ad emi ez pe wee wi **w2**
ae en r **f**
af edn enr exe ln lte rae ren rr **ej fn**
ag emn eoe je wn wte
ah ebe edi ens lee li rei rs **f3 x2**
ai ed ene l re **fe**
aj ekm eno eyt ro **wf**
ak egt ema eq pt wet **wa**
al ece eki end rd rti **em**
am emt eo j wt **wl**
an eg eme p we
ao emm jt wm wtt **ed fe**
ap ekn eye rg rtn **f4 x3**
aq emk **ja** jet wk wnt **ff xn**
ar ec eke enn rn rte **eb f2 xe**
as eb ede eni le ree ri **eb f2 xe**
at em w
au edt ena ex lt ra ret **fa xl xt**
av ebt eda enu la let rit ru **fu xa**
aw ekt enm ey rm rtt **fm**
ax emu ezt pa pet wit wu **fd**
ay egm emw eqt pm ptt wat wem ww **f3 x2**
az emd eoi jee ji wd wne wti **f3 x2**

B

ba dit du neu nst nv tev tht tiu
bb das dle dri nus tvs
bc dar drn nfn nur tvr **d4 n5**
bd dai dl nel nfe nui tvi **dx**
be di nei ns tes tii th **d2 n3 14 t4**
bf dir dsn dve nhn nsr thr **dq**
bg dan dp dwe nep nun tip tvn
bh dis dsi nhi nss ths **d5**
bi die ds nes nh teh the tis **d3 n4 15 t5**
bj dio dum nso nvm tho
bk **daa** drt nft nua tva
bl dfe did dui nsd nvi thd **da nu lv tv**
bm dat dw new nut tiw tvt
bn dae dr nf tef tir tve **df nq**
bo dam dj dwt nej num tij tvm **d3 n4 15**
bp dig dun nsg nvn thg [t5
bq dak dpt dwa nuk tvk
br df din due nef nsn nve thn tif **d4 n5**
bs dh dii dse neh nhe nsi thi tih **d4 n5**
bt da **nea** nu teu tia tv **lv**
bu dia dv nev nht nsa tiv tha
bv diu dht dsa nha nsu thu
bw dim dut nsm nvt thm
bx dau dlt dra nfa nuu tvu
by daw drm nfm nuw tvw **d5**
bz dad dpe dwi nud tvd **d5**

C

ca kit ku ndt ntu nx tau tex tlt **ev 2u 4l**
cb kas kle kri nci nks **v3** [4t
cc kar krn ncn nkr **6**
cd kai kl kre nce nki tab **2x v2**
ce ki nd nti tai ted tl **4**
cf kir ksn kve nbn ndr tbr **4n 2q**
cg kan kp kwe nkn nye tap **vn**
ch khe kis ksi nbi nds tls **7**
ci kie ks nb nde nts tas teb tle **5**
cj kio kum ndo nxm tlo **vf**
ck **kaa** krt nct nka **va**
cl kfe kid kui ndd nxi tld **eu 2a 3t**
cm kat kw nkt ntw ny taw **vl vt 2w**
cn kae kr nc tar **2f eq ve**
co kam kj kwt nkm ntj nyt taj **5**
cp kig kun ndg nxn tlg **8**
cq kak kpt kwa nkk nya **4f 5n**
cr kf kin kue ndn ntf nxe taf tln **6**
cs kh kii kse nbe ndi nth tah tli **6**
ct ka nk nta taa tek **eu 2a 3l**
cu kia kv nbt nda ntv tav tla **2v 4a 5l 5t**
cv kht kiu ksa nba ndu tlu **4u 5a**
cw kim kut ndm nxt tlm **4m**
cx kau klt kra nca nku **4d**
cy kaw krm ncm nkw **7**
cz kad kpe kwi nkd nnz **7**

D

da bt nit nu teu tst tv **bl lv**
db nas nle nri tfi tus xei xie xs
dc nar nrn tfn tur xae xen xr **b2 n4 15 t5**
dd nai nl tfe tui xee xi **nx**
de b ni tei ts **n2 13 t3**
df br ben nir nsn nve thn tsr **bn nq**
dg nan np nwe tep tun xn xte
dh bei bie bs nhe nis nsi thi tss **b3 n5**
di be nie ns tes th tse **n3 14 t4**
dj bmt bo btm nio num tso tvm
dk naa nrt tft xa xet
dl bd bne bti nfe nui tvi **na lu tu**
dm nat nw tew tut xt
dn nr ter tf tue xe **nf lq tq**
do nam nj tej tum xm xtt **be n3 14 t4**
dp bg bme btn nig nun tsg tvn **b4**
dq nak npt nwa tuk xk xnt xta **bf**
dr bn bte nf nin nue tef tsn tve **b2 n4 15 t5**
ds bee bi nh nii nse teh the tsi **b2 n4 15**
dt na tea tu x **lu** [t5
du ba bet nia nv tev tht tsa
dv bea bit bu nht niu nsa tha
dw bm btt nim nut tsm
dx nau nlt nra tfa tuu xea xit xu **bd**
dy naw nrm tfm tuw xat xem xw **b3 n5**
dz nad npe mvi xd xne xti **b3 n5**

E

ea it u **2l 2t**
eb aei aie as le ree ri **a3 f2**
ec aae aen ar rn rte **4**
ed aee ai l re **a2 fe**
ee i
ef iae ien ir sn ste ve **2n ue**
eg an ate p we
eh he iei iie is si **5**
ei ie s **3**
ej imt itm io um utt **af fn**
ek aa aet rt **fl ft**
el fe id ine iti uee ui **a**
em at w **al**
en ae r
eo am att j wt **3**
ep ig ime itn un ute **6**
eq ak ant ata pt wa wet **2f 3n ve**
er f in ite **ue 4**
es h ii se **4**
et a
eu **ia** iet st v **2a 3l 3t**
ev ht iea iit iu sa set **2u 3a 4l 4t**
ew im itt ut **2m ul**
ex aea ait au lt ra ret **qe u2 2d**
ey aat aem aw rm rtt **5**
ez ad ane ati pe wee wi **5**

F

fa eau eex elt idt itu ix uit uu **au xt**
fb ici iks uas ule uri
fc icn ikr uar urn **a4 x2**
fd eal ice iki itl uai ure ul **ax**
fe eai el id iti ui **a2 ed**
ff elr ibn idr uir usn uve **aq xn**
fg eap ikn itp iye uan up uwe
fh els ibi ids uis uhe usi **a5 x3**
fi eas eeb ele ib ide its uie us **a3 eb xe**
fj elo ido ixm uio uum
fk ict ika uaa urt
fl eld idd ixi ufe uid uui **aa ek**
fm eaw eey ikt itw iy uat uw **aw**
fn ear eec ic ike itr uae ur **af aj**
fo eaj ikm itj iyt uam uj uwt **a3 eb xe**
fp elg idg ixn uig uun **x4**
fq ikk iya uak upt uwa **xf**
fr eaf eln idn itf ixe uf uin uue **a4 x2**
fs eah ibe idi eli ith uh uii use **a4 x2**
ft eaa eek ik ita ua **aa ek**
fu eav ela ibt ida itv uia ust uv **av xa**
fv elu iba idu uiu uht usa **xu**
fw elm idm ixt uim uut **xm**
fx ice iku uau ult ura **xd**
fy icm ikw uaw urm **a5 x3**
fz ikd iyi uad upe uwi **a5 x3**

Detector.

G

ga mit mu tdt ttu tx zt
gb mas mle mri qei qie qs tci tks
gc mar mrn qae qen qr tcn tkr **m4**
gd mai ml mre qee qi tce tki ttl **mx**
ge mi td tti z **m2 ld**
gf mir msn tbn tdr zae zen zr **mq**
gg man mp mwe qn qte tkn ttp tye
gh mhe mis msi tbi tds zei zie zs **m5**
gi mie ms tb tde tts ze **m3 lb**
gj mio mum tdo txm zmt zo ztm
gk maa mrt qa qet tct tka
gl mfe mid mui tdd txi zd zne zti **lk ma**
gm mat qt tkt ttw ty **mw** [**tk**
gn mae mr qe tc tke ttr **mf lj tj**
go mam mj mwt qm qtt tkm ttj tyt **lb m3**
gp mig mun tdg txn zg zme ztn [**tb**
gq mak mpt mwa qk qnt qta tkk tya
gr mf min mue tdn ttf zn zte **m4**
gs mh mii mse tbe tdi tth zee zi **m4**
gt ma q tk tta **lk**
gu mia mst mv tbt tda ttv za zet
gv mht miu msa tba tdu zea zit zu
gw mim mut tdm txt zm ztt
gx mau mlt mra qea qit qu tca tku
gy maw mrm qat qem qw tcm tkw **m5**
gz mad mpe mwi qd qne qti tkd tyi **m5**

H

ha eev eht eiu ieu ist sit su **2v 3u 5l 5t**
hb evs ifi ius sas sle sri
hc evr ifa iur sar srn **7**
hd eil evi iel ife iui sai sl sre **3x**
he ees eh eii iei is si **5**
hf ehr ihn isr sir ssn sve **5n 3q**
hg eip evn iep iun san sp swe
hh ehs ihi iss she sis ssi **8**
hi eeh ehe eis ies ise sie ss **6**
hj eho iso ivm sio sum
hk eva ift iua saa srt
hl ehd isd ivi sfe sid sui **ev 2u 3a**
hm eiw evt iew iut sat sw **3w**
hn eef eir eve ier if iue sae sr **3f 2q**
ho eij evm iej ium sam sj swt **6**
hp ehg isg ivn sig sun **9**
hq evk iuk sak spt swa **5f**
hr ehu eif ief isn ive sf sin sue **7**
hs ehi eih ieh ihe isi sh sii **7**
ht eeu eia ev iea iu sa **ev 2u 3a**
hu eha eiv iev iht isa sia sst sv **3v 5a**
hv ehu iha isu sht siu ssa **5u**
hw ehm ism ivt sim sut **5m**
hx evu ifa iuu sau slt sra **5d**
hy evw ifm iuw saw srm **8**
hz evd iud sad spe swi **8**

I

ia eu eit st v **3l 3t**
ib eas ele eri fee fi uei uie us **u3 q2**
ic ear ern fn fte ike uae uen ur **5**
id eai el ere fe uee ui **ex qe u2**
ie ei s **3**
if eir eve hn hte sae sen sr **eq ve 3n**
ig ean ep ewe un ute
ih ehe eis esi hee sie ss **6**
ii eie es h se **4**
ij eio eum smt so stm vm vtt **qn uf**
ik eaa ert ft ua uet **ql qt**
il efe eid eui sd sne sti vee vi **ea ot**
im eat ew ut **ul**
in eae er f ue **ef**
io eam ej ewt um utt **4**
ip eig eun sg sme stn vn vte **7**
iq eak ept ewa uk unt uta **3f 4n**
ir ef ein eue sn ste ve **5**
is eh eii he si **5**
it ea u **ol**
iu eia est ev iea ht sa set **3a 4l 4t**
iv eht eiu esa ha het sea sit su **3u 4a 5l**
iw eim eut sm stt vt **3m** [**5t**
ix eau elt era fa fet uea uit uu **v2 3d**
iy eaw erm fm ftt uat uem uw **6**
iz ead epe ewi ud une uti **6**

J

ja agt aq atr emk etq wk wnt **ku**
jb aos wgi wms wze
jc aor wgn wmr wqe **k4**
jd aoi atz emz wge wmi wz **kx**
je ag atn emn etg wn **k2 lx nd tx**
jf agr azn wdn wnr wxe **kq**
jg aon wmn woe
jh ags azi wbe wdi wns **kp**
ji age atd az emd etz wd wne **k3 nb**
jj ago aqm wkm wno wyt
jk aoa atq emq wgt wma wq
jl agd aqi wce wki wnd **ka nk**
jm aot ato emo wmt wo **kw**
jn aoe atg emg wg wme **kf nj**
jo aom wmm wot **k3 nb**
jp agg aqn wkn wng wye
jq aok wmk woa
jr agn aqe atc emc wc wke wnn **k4**
js agi atb aze emb wb wde wni **k4**
jt oa atm emm eto wm **ka nk**
ju aga atx azt emx wdt wna wx **kv**
jv agu aza wbt wda wnu
jw agm aqt aty emy wkt wnm wy
jx aou wga wmu wzt
jy aow wgm wmw wqt **k5**
jz aod wmd woi **k5**

K

ka ct nk nnt tek trt **jl jt**
kb ngi nms nze tpi tws yei yie ys
kc ngn nmr nqe tpn twr yae yen yr **j2 nb**
kd nge nmi nz tez tpe twi yee yi
ke c nn ten tr **lf tf**
kf cae cen cr ndn nnr nxe tln trr **jn nj**
kg nmn noe tje twn yn yte
kh cei cie cs nbe ndi nns tli trs **j3**
ki ce nd nne ted tl tre **je lx tx**
kj cmt co ctm nkm nno nyt tro
kk ngt nma nq teq tpt twa ya yet
kl cd cne cti nce nki nnd trd **nm lw tw**
km nmt no tj twt yt
kn ng nme teg tp twe ye
ko nmm not tjt twm ym ytt **je nd lx tx**
kp cg cme ctn nkn nng nye trg **j4**
kq nmk noa tja twk yk ynt yta **jf**
kr cn cte nc nke nnn tec trn **j2 nb**
ks cee ci nb nde nni teb tle tri **j2 nb**
kt nm tem y **lw tw**
ku ca cet ndt nna nx tex tlt tra **ja**
kv cea cit cu nbt nda nnu tla tru **ju**
kw cm ctt nkt nnm ny tey trm **jm**
kx nga nmu nzt tpa twu yea yit yu **jd**
ky ngm nmw nqt tpm tww yat yem yw **j3**
kz nmd noi tji twd yd yne yti **j3**

L

la aeu ast av ebt enu etv rit ru **nl nt**
lb afi aus exs ras rle rri **g2 m3**
lc afn aur exr rar rrn **b de n2 t3**
ld ael afe aui enl exi rai rl rre **ge m2**
le aei as eb eni ets ri
lf ahn asr ebr rir rsn rve **ke nn**
lg aep aun enp exn ran rp rwe **mn**
lh ahi ass ebs lse rhe ris rsi **be d2 n3 t4**
li aes ah ase ebe ens eth lee rie rs **ne to**
lj aso avm ebo rio rum **gn mf**
lk aft aua exa raa rrt **gl gt ma**
ll asd avi ebd rfe rid rui **tt**
lm aew aut enw ext rat rw **ml mt**
ln aer af aue enr etf exe rae rr **me**
lo aej aum enj exm ram rj rwt **ne ti**
lp asg avn ebg rig run **b2 d3 n4**
lq auk exk rak rpt rwa **dn nf**
lr aef asn ave ebn enf rf rin rue b **de n2 t3**
ls aeh ahe asi ebi enh rh rii rse b **de n2 t3**
lt aea au ena etu ex ra **tl**
lu aev aht asa eba env ria rst rv **dl dt na**
lv aha asu ebu rht riu rsa **bl bt da nu**
lw asm avt ebm rim rut **kl kt nm**
lx afa auu exu rau rlt rra **je k2 nd**
ly afm auw exw raw rrm **be d2 n3 t4**
lz aud exd rad rpe rwi **be d2 n3 t4**

[redacted]

[redacted]

Detector.

M

ma gt q tk tnt **lk**
mb oei oie os tgi tms tze
mc oae oen or tgn tmr tqe **g2 lb tb**
md oee oi tge tmi tz
me g tn **ln**
mf gae gen gr tdn tnr txe zn zte **gn lj tj**
mg on ote tmn toe
mh gei gie gs tbe tdi tns zee zi **g3**
mi ge td tne z **ld**
mj gmt go gtm qtt tkm tno tyt
mk oa oet tgt tma tq
ml gd gne gti qee qi tce tki tnd **lm tm**
mm ot to
mn oe tg tme **lg**
mo om ott tmm tot **ge ld td**
mp gg gme gtn qn qte tkn tng tye **g4**
mq ok ont ota tmk toa **gf**
mr gn gte qe tc tke tnn **g2 lb tb**
ms gee gi tb tde tni ze **g2 lb tb**
mt o tm **lm**
mu ga get tdt tna tx zt
mv gea git gu tbt tda tnu za zet
mw gm gtt qt tkt tnm ty
mx oea oit ou tga tmu tzt **gd**
my oat oem ow tgm tmw tqt **g3**
mz od one oti tmd toi **g3**

N

na dt tit tu x **dl lu**
nb cee ci kei kie ks tas tle tri **j2 k3**
nc cn cte kae ken kr tar trn **be d2 l4 t4**
nd ce kee ki tai tl tre **je k2 lx tx**
ne d ti **li lo to**
nf bn bte dae den dr tir tsn tve **dn lq tq**
ng kn kte tan tp twe ye
nh bee bi dei die ds the tis tsi **b2 d3 l5** [**t5**
ni b de tie ts **l3 t3**
nj dmt do dtm tio tum xm xtt **jn kf**
nk ct ka ket taa trt **jl jt**
nl dd dne dti tfe tid tui xee **xi la ta**
nm kt tat tw y **lw kl**
nn c ke tae tr **lf tf**
no km ktt tam tj twt yt **de l3 t3**
np dg dme dtn tig tun xn xte **b3 d4**
nq kk knt kta tak tpt twa ya yet **bn df**
nr dn dte tf tin tue xe **be d2 l4 t4**
ns be dee di th tii tse **be d2 l4 t4**
nt k ta **la**
nu bt da det tia tst tv **bl lv**
nv ba bet dea dit du tht tiu tsa
nw dm dtt tim tut xt
nx ca cet kea kit ku tau tlt tra **dd**
ny cm ctt kat kem kw taw trm **b2 d3 l5 t5**
nz kd kne kti tad tpe twi yee yi **b2 d3 l5 t5**

O

oa mk mnt tgt tq ttk **eu 3l 3t**
ob mgi mms mze ode tos **q2 u3**
oc mgn mmr mqe tor **5**
od mge mmi mz toi ttz **ex qe u2**
oe mn tg ttn **3**
of mdn mnr mxe tgr tzn **eq ve 3n**
og mmn moe ton **un**
oh mbe mdi mns tgs tzi **6**
oi md mne tge ttd tz **4**
oj mkm mno myt tgo tqm **qn uf**
ok mgt mma mq toa ttq **ql qt ua**
ol mce mki mnd tgd tqi **ea it**
om mmt mo tot tto **ew ul ut**
on mg mme toe ttg **ef ue**
oo mmm mot tom **4**
op mkn mng mye tgg tqn **7**
oq mmk moa tok **3f 4n**
or mc mke mnn tgn tqe ttc **5**
os mb mde mni tgi ttb tze **5**
ot mm to ttm **ea il**
ou mdt mna mx tga ttx tzt **ev 3a 4l 4t**
ov mbt mda mnu tgu tza **3u 4a 5l 5t**
ow mkt mnm my tgm tqt tty **3m**
ox mga mmu mzt odt tou **v2 3d**
oy mgm mmw mqt tow **6**
oz mmd moi tod **6**

P

pa adt atu ax emu etx ezt wit wu **3v 4u**
pb aci aks eqs was wle wri
pc acn akr eqr war wrn **8**
pd ace aki atl eml eqi wai wl wre **4x**
pe ad ati emi etd ez wi **6**
pf abn adr ezr wir wsn wve **4q**
pg akn atp aye emp eqn wan wp wwe
ph abi ads ezs whe wis wsi **9**
pi ab ade ats ems etb eze wie ws **7**
pj ado axm ezo wio wum
pk act aka eqa waa wrt
pl add axi ezd wfe wid wui **2v 3u 4a**
pm akt atw ay emw eqt ety wat ww **4w**
pn ac ake atr emr eqe etc wae wr **3q 4f**
po akm atj ayt emj eqm wam wj wwt **7**
pp adg axn ezg wig wun
pq akk aya eqk wak wpt wwa
pr adn atf axe emf ezn wf win wue **8**
ps abe adi ath emh ezi wh wii wse **8**
pt ak ata ema eq etk wa **2v 3u 4a**
pu abt ada atv emv eza wia wst wv **4v**
pv aba adu ezu wht wiu wsa
pw adm axt ezm wim wut
px aca aku equ wau wlt wra
py acm akw eqw waw wrm **9**
pz akd ayi eqd wad wpe wwi **9**

Q

qa gk gnt maa mek mrt tct tnk
qb ggi gms gze mab mpi mws **tys**
qc ggn gmr gqe mpn mwr tyr **u4**
qd gge gmi gz mez mpe mwi tnz tyi **ux**
qe gn men mr tc tnn ttr **2d ex u2**
qf gdn gnr gxe mln mrr tcr **uq**
qg gmn goe mje mwn tyn
qh gbe gdi gns mli mrs tcs **up**
qi gd gne med ml mre tce tnd ttl **2b u3**
qj gkm gno gyt mro tco
qk ggt gma gq meq mpt mwa tnq tya
ql gce gki gnd mrd.tcd **2k ua**
qm gmt go meo mj mwt tno ttj tyt **uw**
qn gg gme meg mp mwe tng ttp tye **2j uf**
qo gmm got mjt mwm tym **2b u3**
qp gkn gng gye mrg tcg
qq gmk goa mja mwk tyk
qr gc gke gnn mec mrn tcn tnc **u4**
qs gb gde gni meb mle mri tci tnb **u4**
qt gm mem mw tnm ttw ty **2k ua**
qu gdt gna gx mex mlt mra tca tnx **uv**
qv gbt gda gnu mla mru qst tcu
qw gkt gnm gy mey mrm tcm tny
qx gga gmu gzt mpa mwu tyu
qy ggm gmw gqt mpm mww tyw **up**
qz gmd goi mji mwd tyd **up**

R

ra ait au edt etu ex lt **ev 2u 4l 4t**
rb aas ale ari eci eks **v3**
rc aar aru ecn ekr **6**
rd aai al are ece eki etl **v2 2x**
re ai ed eti l **4**
rf air asn ave ebn edr lae len lr **2q 4n**
rg aan ap awe ekn etp eye **vn**
rh ahe ais asi ebi eds lei lie ls **7**
ri aie as eb ede ets le **5**
rj aio aum edo exm lmt lo ltm **vf**
rk aaa art ect eka **va**
rl afe aid aui edd exi ld lne lti **eu 2a 3t**
rm aat aw ekt etw ey **vl vt 2w**
rn aae ar ec eke etr **eq ve 2f**
ro aam aj awt ekm etj eyt **5**
rp aig aun edg exn lg lme ltn **8**
rq aak apt awa ekk eya **pn 4f**
rr af ain aue edn etf exe ln lte **6**
rs ah aii ase ebe edi eth lee li **6**
rt aa ek eta **eu 2a 3l**
ru aia ast av ebt eda etv la let **2v 4a 5l 5t**
rv aht aiu asa eba edu lea lit lu **4u 5a**
rw aim aut edm ext lm ltt **4m**
rx aau alt ara eca eku **4d**
ry aaw arm ecm ekw **7**
rz aad ape awi ekd eyi **7**

S

sa eeu est ev ht iit iu **2u 4l 4t**
sb efi eus ias ile irí vei vie vs **v3**
sc efn eur iar irn vae ven vr **6**
sd eel efe eui iai il vee vi **v2 2x**
se eei es h ii **4**
sf ehn esr hae hen hr iir isn ive **2q 4n**
sg eep eun ian ip iwe vn vte
sh ehi ess hei hie hs ihe isi **7**
si ees eh ese he iie is **5**
sj eso evm hmt ho htm iio ium **vf**
sk eft eua iaa irt va vet
sl esd evi hd hne hti ife iid iui **eu v 2a 3t**
sm eew eut iat iw vt **vl 2w**
sn eer ef eue iae ir ve **eq 2f**
so eej eum iam ij iwt vm vtt **5**
sp esg evn hg hme htn iig iun **8**
sq euk iak ipt iwa vk vnt vta **pn 4f**
sr eef esn eve hn hte if iin iue **6**
ss eeh ehe esi hee hi ih iii ise **6**
st eea eu ia v **2a 31**
su eev eht esa ha het iia ist iv **2v 4a 5l 5t**
sv eha esu hea hit hu iht iiu isa **4u 5a**
sw esm evt hm htt iim iut **4**m
sx efa euu iau ilt ira vea vit vu **4**d
sy efm euw iaw irm vat vem vw **7**
sz eud iad ipe iwi vd vne vti **7**

T

ta k tet **nl nt**
tb gee gi mei mie ms ze **g2 m3**
tc gn gte mae men mr qe **de 13 n2**
td ge mee mi z **m2**
te n
tf dn dte nae nen nr xe **ke nn**
tg mn mte oe
th be dee di nei nie ns **d2 14 n3**
ti d ne **lo**
tj km ktt nmt no ntm yt **gn mf**
tk gt ma met q **gl**
tl ce kee ki nd nne nti **lt**
tm mt o **ml**
tn g me
to mm mtt ot **ne li**
tp kn kte ng nme ntn ye **b2 d3 n4**
tq mk mnt mta oa oet **dn nf**
tr c ke nn nte **de 13 n2**
ts b de nee ni **de 13 n2**
tt m **ll**
tu dt na net **x**
tv bt da det nea nit nu **bl**
tw kt nm ntt y **kl**
tx ga get mea mit mu zt **je k2 nd**
ty gm gtt mat mem mw qt **be d2 14 n3**
tz md mne mti oee oi **be d2 14 n3**

U

ua eek ert ft ik int **2k ql qt**
ub epi ews igi ims ize
uc epn ewr ign imr iqe **q2 2**b
ud eez epe ewi ige imi iz
ue een er f in **ef 2n**
uf eln err fr idn inr ixe **2j qn**
ug eje ewn imn ioe
uh eli ers fei fie fs ibe idi ins **q3**
ui eed el ere fe id ine **ex 2d qe**
uj ero fmt fo ftm ikm ino iyt
uk eeq ept ewa igt ima iq
ul erd fd fne fti ice iki ind **ew 2m**
um eeo ej ewt imt io
un eeg ep ewe ig ime **2g**
uo ejt ewm imm iot **ex 2d qe**
up erg fg fme ftn ikn ing iye **q4**
uq eja ewk imk ioa **qf**
ur eec ern fn fte ic ike inn **q2 2b**
us eeb ele eri fee fi ib ide ini **q2 2b**
ut eem ew im **2**m
uu ecx elt era fa fet idt ina ix
uv ela eru fea fit fu ibt ida inu **qu**
uw eey erm fm fit ikt inm iy **qm**
ux epa ewu iga imu izt **qd**
uy epm eww igm imw iqt **q3**
uz eji ewd imd ioi **q3**

V

va eft eik iek irt sk snt **3k**
vb ipi iws sgi sms sze vde
vc ipn iwr sgn smr sqe vke **3b**
vd eiz iez ipe iwi sge smi sz
ve eer ef ein ien ir sn **2f 3n eq**
vf efr iln irr sdn snr sxe **3j**
vg ije iwn smn soe
vh efs ili irs sbe sdi sns
vi eel efe eid ied il ire sd sne **2x 3d**
vj efo iro skm sno syt
vk eiq ieq ipt iwa sgt sma sq
vl efd ird sce ski snd **2w 3m**
vm eej eio ieo ij iwt smt so
vn eep eig ieg ip iwe sg sme **3g**
vo ijt iwm smm sot **2x 3d**
vp efg irg skn sng sye
vq ija iwk smk soa
vr efn eic iec irn sc ske snn **3b**
vs efi eib ieb ile iri sb sde sni **3b**
vt eew eim iem iw sm **2w 3m**
vu efa eix iex ilt ira sdt sna sx
vv efu ila iru sbt sda snu
vw efm eiy iey irm skt snm sy
vx ipa iwu sga smu stx szt
vy ipm iww sgm smw sqt
vz iji iwd smd soi

W

wa ak ant egt eq etk pt
wb agi ams aze eos jei jie js
wc agn amr aqe eor jae jen jr **ab**
wd age ami az eoi etz jee ji
we an eg etn p
wf adn anr axe egr ezn pae pen pr **aj**
wg amn aoe eon jn jte
wh abe adi ans egs ezi pei pie ps
wi ad ane ege etd ez pe
wj akm ano ayt ego eqm pmt po ptm
wk agt ama aq eoa etq ja jet
wl ace aki and egd eqi pd pne pti **am**
wm amt ao eot eto jt
wn ag ame eoe etg je
wo amm aot eom jm jtt **ad**
wp akn ang aye egg eqn pg pme ptn
wq amk aoa eok jk jnt jta
wr ac ake ann egn eqe etc pn pte **ab**
ws ab ade ani egi etb eze pee pi
wt am eo etm j
wu adt ana ax ega etx ezt pa pet
wv abt ada anu egu eza pea pit pu
ww akt anm ay egm eqt ety pm ptt
wx aga amu azt eou jea jit ju
wy agm amw aqt eow jat jem jw
wz amd aoi eod jd jne jti

X

xa dk dnt nek nrt tft tik **av**
xb dgi dms dze npi nws
xc dgn dmr dqe npn nwr **ap f4**
xd dge dmi dz nez npe nwi tiz **fx**
xe dn nen nr ter tf tin **a3 eb f2**
xf ddn dnr dxe nln nrr tfr **fq**
xg dmn doe nje nwn
xh dbe ddi dns nli nrs tfs **f5**
xi dd dne ned nl nre tel tfe tid **a4 f3**
xj dkm dno dyt nro tfo
xk dgt dma dq neq npt nwa tiq
xl dce dki dnd nrd tfd **au fa**
xm dmt do neo nj nwt tej tio **fw**
xn dg dme neg np nwe tep tig **ff aq**
xo dmm dot njt nwm **a4 f3**
xp dkn dng dye nrg tfg
xq dmk doa nja nwk
xr dc dke dnn nec nrn tfn tic **a5 f4**
xs db dde dni neb nle nri tfi tib **a5 f4**
xt dm nem nw tew tim **au fa**
xu ddt dna dx nex nlt nra tfa tix **fv**
xv dbt dda dnu nla nru tfu
xw dkt dnm dy ney nrm tfm tiy
xx dga dmu dzt npa nwu
xy dgm dmw dqt npm nww **f5**
xz dmd doi nji nwd **f5**

Detector.

Y

ya kk knt ngt nq ntk tak teq tpt **2v 3u 51**
yb kgi kms kze nos tjs [**5t**
yc kgn kmr kqe nor tjr **7**
yd kge kmi kz noi ntz taz tji **3x**
ye kn ng ntn tan teg tp **5**
yf kdn knr kxe ngr nzn tpr **pn 3q**
yg kmn koe non tjn
yh kbe kdi kns ngs nzi tps **8**
yi kd kne nge ntd nz tad tez tpe **6**
yj kkm kno kyt ngo nqm tpo
yk kgt kma kq noa ntq taq tja
yl kce kki knd ngd nqi tpd **ev 2u 3a**
ym kmt ko not nto tao tjt **3w**
yn kg kme noe ntg tag tje **2q 3f**
yo kmm kot nom tjm **6**
yp kkn kng kye ngg nqn tpg **9**
yq kmk koa nok tjk **5f**
yr kc kke knn ngn nqe ntc tac tpn **7**
ys kb kde kni ngi ntb nze tab tpi **7**
yt km no ntm tam teo tj **ev 2u 3a**
yu kdt kna kx nga ntx nzt tax tpa **3v 5a**
yv kbt kda knu ngu nza tpu **5u**
yw kkt knm ky ngm nqt nty tay tpm **5m**
yx kga kmu kzt nou tju **5d**
yy kgm kmw kqt now tjw **8**
yz kmd koi nod tjd **8**

Z

za git gu meu mst mv tbt tnu ttv **2v 3u 51**
zb gas gle gri mfi mus txs [**5t**
zc gar grn mfn mur txr **7**
zd gai gl gre mel mfe mui tnl txi **3x**
ze gi mei ms tb tni tts **5**
zf gir gsn gve mhn msr tbr **3q 5n**
zg gan gp gwe mep mun tnp txn
zh ghe gis gsi mhi mss tbs **8**
zi gie gs mes mh mse tbe tns tth **6**
zj gio gum mso mvm tbo
zk gaa grt mft mua txa
zl gfe gid gui msd mvi tbd **ev 2u 3a**
zm gat gw mew mut tnw txt **3w**
zn gae gr mer mf mue tnr ttf txe **3f 2q**
zo gam gj gwt mej mum tnj txm **6**
zp gig gun msg mvn tbg **9**
zq gak gpt gwa muk txk **5f**
zr gf gin gue mef msn mve tbn tnf **7**
zs gh gii meh mhe msi tbi tnh **7**
zt ga mea mu tna ttu tx **ev 2u 3a**
zu gia gst gv mev mht msa tba tnv **3v 5a**
zv ght giu gsa mha msu tbu **5u**
zw gim gut msm mvt tbm **5m**
zx gau glt gra mfa muu txu **5d**
zy gaw grm mfm muw txw **8**
zz gad gpe gwi mud txd **8**

European.

		Wrong Spacing	Drop First Signal	Drop Last Signal	Change . to —	Change — to .
.—	**a**	et	t	e	m	i
—...	**b**	de ni ts	s	d	c x z	h
—.—.	**c**	ke nn ten tr	r	k	y	b f
—..	**d**	ne ti	i	n	g k	s
.	**e**				t	
..—.	**f**	er in ue	r	u	c p	h
——.	**g**	me tn	n	m	o	d r
....	**h**	es ii se	s	s	b f l v	
..	**i**	ee	e	e	a n	
.———	**j**	{ am att emt / etm eo wt }	o	w		p
—.—	**k**	nt ta tet	a	n	o	d u
.—..	**l**	ai ed re	d	r	p z	h
——	**m**	tt	t	t		a n
—.	**n**	te	e	t	m	i
———	**o**	mt tm	m	m		g k w
.——.	**p**	an ate eg we	g	w	j	f l
——.—	**q**	gt ma tk	k	g		x z
.—.	**r**	ae en	n	a	g w	s
...	**s**	ie ei	i	i	d r u	
—	**t**					e
..—	**u**	ea it	a	i	k w	s
...—	**v**	eu ia st	u	s	x	h
.——	**w**	at em	m	a	o	r u
—..—	**x**	dt na tu	u	d	q y	b v
—.——	**y**	kt nm tw	w	k		c x
——..	**z**	ge mi td	d	g	q	b l

Some letters can be converted into 2 by the double error of change of a signal and wrong spacing. This rarely occurs and a mutilation of that character can only be found by a strict analysis of the signals and possible conversion. Thus "c" might become "oe" by changing first . to — and wrong spacing the final .

In the same way "q" might become "mm," "aa," "ot," "na,"letc.

Detector.

American.

		Wrong Spacing	Drop First Signal	Drop Last Signal	Change . to —	Change — to .
.—	a	el et	l t	e	m	**2**
—...	b	de l3 n2 t3	**3**	d	j	**4**
.. .	c	**3**	**2**	**2**	d f u	
—..	d	l2 ne t2	**2**	n	g k	
.	e				t	
.—.	f	ae en	n	a	g w	**3**
——.	g	me ln tn	n	m		d f
....	h	**4**	**3**	**3**	b q v x	
..	i	**2**	e	e	a n	
—.—.	j	{ ke len lf nn ten tf }	f	k		b q
—.—	k	{ la lel let nl tel tet }	a	n		d u
——	l	t				e
——	m	ll lt tl tt	l t	l t		a n
—.	n	le te	e	l t	m	**2**
. .	o	**2**	e	e	a n	
.....	p	**5**	**4**	**4**		
..—.	q	ef 2n ue	f	u	j	**4**
. ..	r	**3**	**2**	**2**	d f u	
...	s	**3**	**2**	**2**	d f u	
—	t	l				e
..—	u	ea 2l 2t	a	**2**	k w	**3**
...—	v	eu 2a 3l 3t	u	**3**		**4**
.——	w	al at em	m	a		f u
.—..	x	a2 ed fe	d	f		**4**
.. ..	y	**4**	**3**	**3**	b q v x	
... .	z	**4**	**3**	**3**	b q v x	

[redacted]

[redacted]

Detector.

Signal Conversion Table.

.	E	
..	I	Am. O
...	S	Am. C or R
....	H	Am. Y or Z
.—	A	
.—.	R	Am. F
.—..	L	Am. X
.——	W	
.———	J	
.——.	P	
..—	U	
..—.	F	Am. Q
...—	V	
—	T	Am. L
—.	N	
—..	D	
—...	B	
—.—	K	
—.—.	C	Am. J
—.——	Y	
—..—	X	
——	M	
——.	G	
——.—	Q	
——..	Z	
———	O	

This Table will assist in discovering conversions of signals quickly.

Space or ("dot") Letters.

2	ee i o
3	c ei eo ie oe r s
4	ce ec er es h ii io oi oo re se y z
5	ci co eh ey ez he ic ir is oc or os p ri ro si so ye ze
6	cc cr cs ep hi ho ih iy iz oh oy oz pe rc rr rs sc sr ss yi yo zi zo
7	ch cy cz hc hr hs ip op pi po rh ry rz sh sy sz yc yr ys zc zr zs
8	cp hh hy hz pc pr ps rp sp yh yy yz zh zy zz
9	hp ph py pz yp zp

NOTE.

American space letters are c . . . e . h . . . . i . . o . . p . . . . r . . . s . . . y . . . . z . . . .

European space letter are e . h . . . . i . . s . . .

SECTIONS.

A A=0

A B=1

B B=2

B A=3

www.ingramcontent.com/pod-product-compliance
Lightning Source LLC
LaVergne TN
LVHW062306040326
832903LV00009B/268
9781608883141